ATLAS
OF
THE
SKELETAL
MUSCLES

ATLAS OF THE SKELETAL MUSCLES

Robert J. Stone
Suffolk Community College

Judith A. Stone
Suffolk Community College

 WCB   **Wm. C. Brown Publishers**

Book Team

Editor *Edward G. Jaffe*
Project Editor *Colin H. Wheatley*
Production Coordinator *Carla D. Arnold*

 Wm. C. Brown Publishers

President *G. Franklin Lewis*
Vice President, Publisher *George Wm. Bergquist*
Vice President, Publisher *Thomas E. Doran*
Vice President, Operations and Production *Beverly Kolz*
National Sales Manager *Virginia S. Moffat*
Advertising Manager *Ann M. Knepper*
Marketing Manager *Craig S. Marty*
Executive Editor *Edward G. Jaffe*
Production Editorial Manager *Colleen A. Yonda*
Production Editorial Manager *Julie A. Kennedy*
Publishing Services Manager *Karen J. Slaght*
Manager of Visuals and Design *Faye M. Schilling*

Cover design by Jeanne Marie Regan

Library of Congress Catalog Card Number: 89–61734

ISBN 0–697–10618–7

Printed in the United States of America by Wm. C. Brown Publishers, 2460 Kerper Boulevard, Dubuque, IA 52001

10 9 8

DEDICATION

To Karen, Andrew, and Laura for their interest,
enthusiasm, and cooperation.

CONTENTS

CHAPTER FOUR MUSCLES OF THE NECK 49

CHAPTER FIVE MUSCLES OF THE TRUNK 71

CHAPTER SIX MUSCLES OF THE SHOULDER AND ARM 97

CHAPTER SEVEN MUSCLES OF THE FOREARM AND HAND 119

PREFACE

This book is a study guide to the anatomy and actions of human skeletal muscles. It is designed for use by students of physical therapy, chiropractic, medicine, nursing, physical education, and other health-related fields. It also serves as a compact reference for the practicing professional.

The first chapter presents labeled line drawings of the skeleton, which include all structures that are used in describing origins and insertions in the later chapters. A master numbering system is used so that each structure is labeled with the same number in all drawings.

The second chapter describes the various movements of the body.

In chapters three through nine the origin, insertion, action, and innervation of the skeletal muscles are described and each muscle is presented on a separate page with a line drawing. The spinal cord level of the nerve fibers that innervate each muscle are included in parentheses after the name of each nerve.

The drawings include the following important features:

1. Bones and cartilage containing muscle attachment sites are shaded
2. Adjacent structures are shown
3. Muscle fibers are drawn by direction
4. Muscle fibers are shown on the undersurface of bone and cartilage as dashed lines
5. Tendons and aponeuroses are shown

These features aid in visual orientation and understanding of the action of the muscles.

Since our primary goal is to describe the muscles moving the skeleton we have not described the muscles of the perineum, eye, tympanic cavity, tongue, larynx, pharynx, or palate.

We extend our appreciation to Mr. George Boykin, the jolly proprietor of the gross anatomy laboratories at the Health Sciences Center of the State University of New York at Stony Brook for his help and encouragement over the years. We also thank Mr. Vincent Verdisco and Ms. Diane Chandler for their technical advice with the artwork, Ms. Katherine Juner for her secretarial services, and the following list of reviewers: S. R Peterson, University of Alberta; Sherwin Mizell, Indiana University; Danielle Desroches, William Paterson College; Michael R. Hawes, The University of Calgary; Farica R. Bialstock, Nassau Community College; and Robert J. Gurney, Grant MacEwan College.

Robert J. Stone

Judith A. Stone

CHAPTER ONE
THE SKELETON

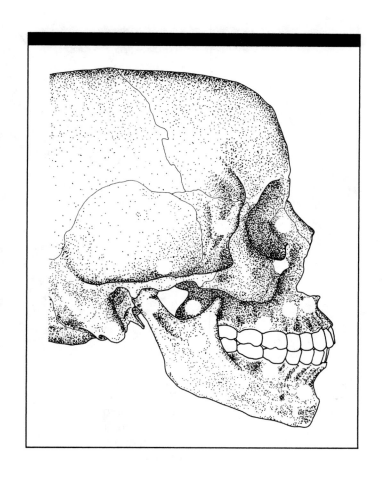

BONES OF THE SKULL—LATERAL VIEW

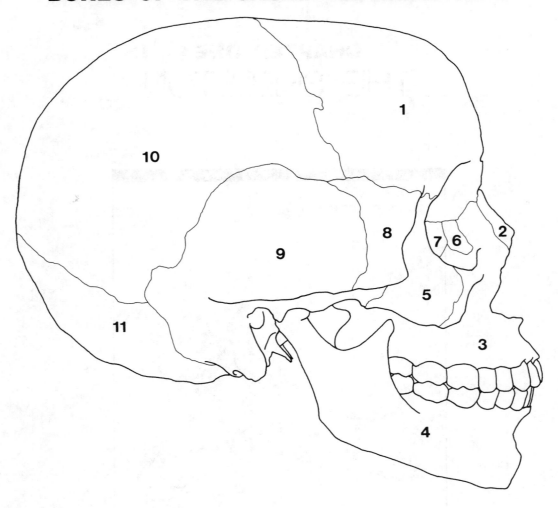

1. Frontal
2. Nasal
3. Maxilla
4. Mandible
5. Zygomatic

6. Lacrimal
7. Ethmoid
8. Sphenoid
9. Temporal
10. Parietal
11. Occipital

SKULL—LATERAL VIEW

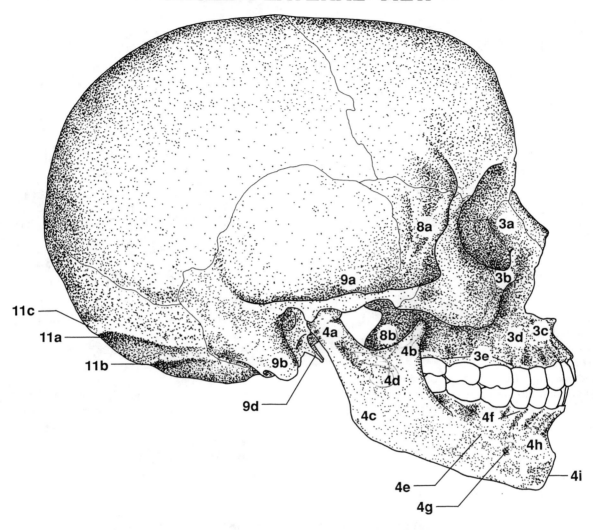

3a.	Frontal process (maxilla)	**4g.**	Mental foramen (mandible)
3b.	Zygomatic process (maxilla)	**4h.**	Incisive fossa of mandible
3c.	Incisive fossa of maxilla	**4i.**	Symphysis of mandible
3d.	Canine fossa (maxilla)	**8a.**	Greater wing of sphenoid bone
3e.	Alveolar border of maxilla	**8b.**	Lateral pterygoid plate
4a.	Neck of condyle (mandible)	**9a.**	Temporal fossa
4b.	Coronoid process (mandible)	**9b.**	Mastoid process (temporal bone)
4c.	Angle of the mandible	**9d.**	Styloid process (temporal bone)
4d.	Ramus (mandible)	**11a.**	Superior nuchal line (occipital bone)
4e.	Oblique line (mandible)	**11b.**	Inferior nuchal line (occipital bone)
4f.	Alveolar border of mandible	**11c.**	External occipital protuberance

SKULL—LATERAL VIEW

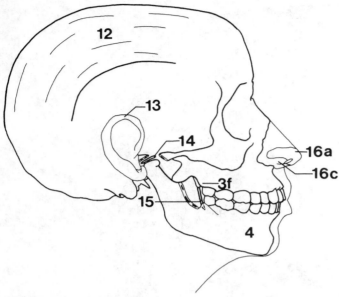

3f. Tuberosity of maxilla
4. Mandible
12. Galea aponeurotica
13. Helix of ear
14. Articular disk of temporomandibular joint
15. Pterygomandibular raphe
16a. Greater alar cartilage
16c. Ala

SKULL—POSTERIOR VIEW

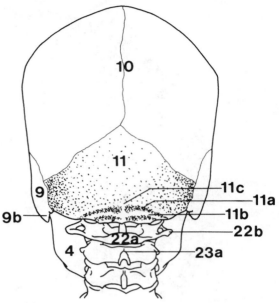

4. Mandible
9. Temporal bone
9b. Mastoid process (temporal bone)
10. Parietal bone
11. Occipital bone
11a. Superior nuchal line (occipital bone)
11b. Inferior nuchal line (occipital bone)
11c. External occipital protuberance
22a. Posterior arch of atlas
22b. Transverse process of atlas
23a. Spinous process of axis

SKULL—ANTERIOR VIEW

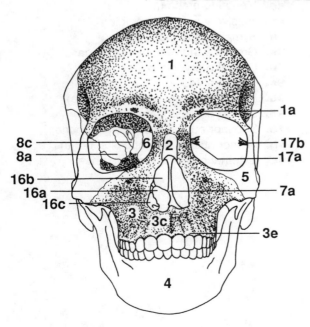

1. Frontal bone
1a. Superciliary arch (frontal bone)
2. Nasal bone
3. Maxilla
3c. Incisive fossa of maxilla
3e. Alveolar border of maxilla
4. Mandible
5. Zygomatic bone
6. Lacrimal bone
7a. Nasal septum
8a. Greater wing of sphenoid bone
8c. Lesser wing of sphenoid bone
16a. Greater alar cartilage
16b. Nasal cartilage
16c. Ala
17a. Medial palpebral ligament
17b. Lateral palpebral raphe (ligament)

SKULL TO HUMERUS—LATERAL VIEW

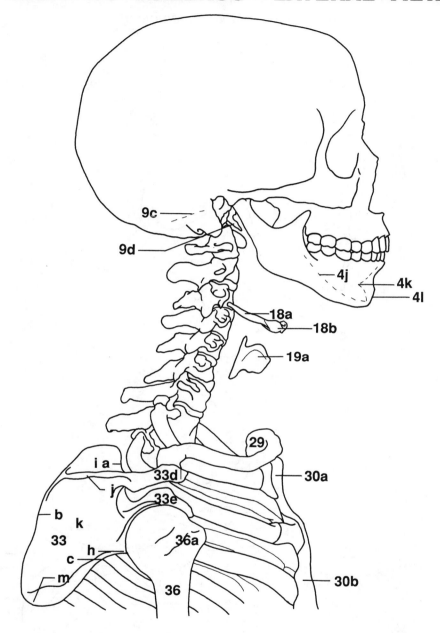

4j.	Mylohyoid line (medial surface of mandible)	**33a.**	Superior border of scapula
4k.	Inferior mental spine (inner surface of mandible)	**33b.**	Vertebral (medial) border of scapula
4l.	Symphysis of mandible	**33c.**	Axillary (lateral) border of scapula
9c.	Mastoid notch (medial surface of temporal bone)	**33d.**	Acromion (scapula)
9d.	Styloid process (temporal bone)	**33e.**	Coracoid process (scapula)
18a.	Greater cornu of hyoid	**33h.**	Infraglenoid tubercle (scapula)
18b.	Body of hyoid	**33i.**	Supraspinous fossa (scapula)
19a.	Lamina of thyroid cartilage	**33j.**	Crest of spine (scapula)
29.	Clavicle	**33k.**	Infraspinous fossa (scapula)
30a.	Manubrium	**33m.**	Inferior angle of scapula
30b.	Body of sternum	**36.**	Humerus
33.	Scapula	**36a.**	Greater tuberosity of humerus

SKULL TO STERNUM—ANTERIOR VIEW

(Mandible and maxilla removed)

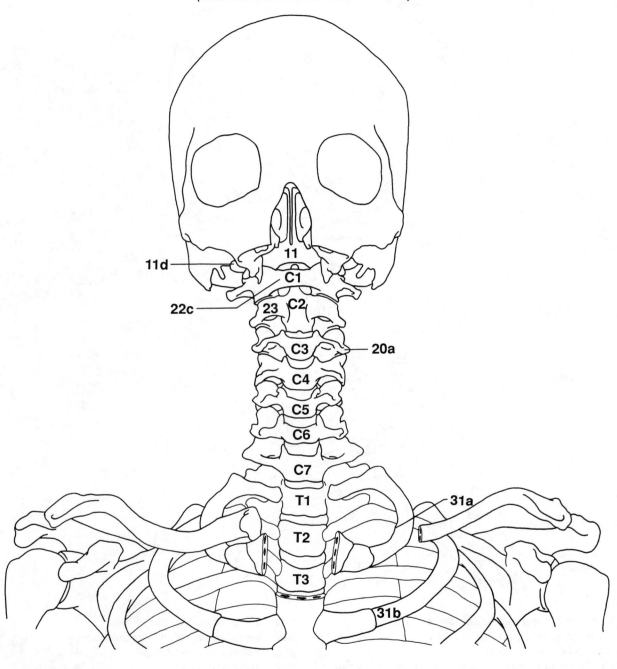

11. Occipital bone	**22c.** Anterior arch of atlas
11d. Jugular process of occipital bone	**23.** Axis
20a. Transverse process of vertebra	**31a.** Scalene tubercle of first rib
	31b. Second rib

RIB CAGE, PELVIC GIRDLE, UPPER ARM—ANTERIOR VIEW

(Ribs partially removed, right arm disarticulated)

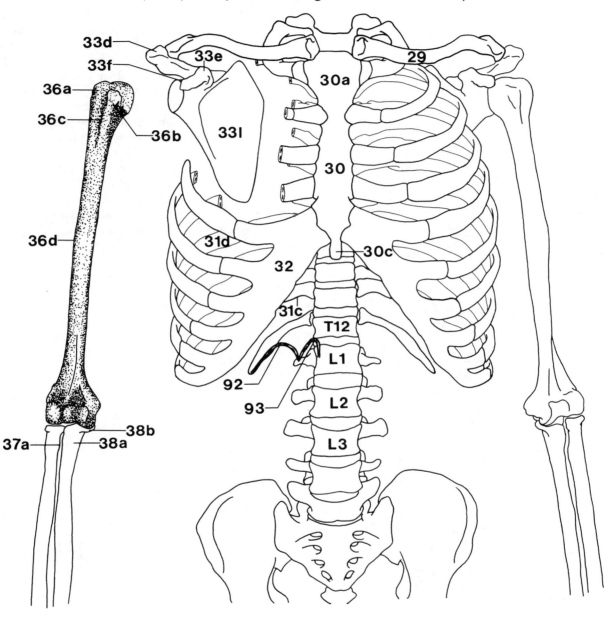

29. Clavicle	**33l.** Subscapular fossa (scapula)
30. Sternum	**36.** Humerus
30a. Manubrium	**36a.** Greater tuberosity (tubercle) of the humerus
30c. Xiphoid process	**36b.** Lesser tuberosity of the humerus
31c. Tubercle of rib	**36c.** Intertubercular (bicipital) groove (humerus)
31d. Angle of rib	**36d.** Deltoid tuberosity (humerus)
32. Costal cartilage	**37a.** Radial tuberosity (radius)
33. Scapula	**38a.** Ulnar tuberosity (ulna)
33d. Acromion (scapula)	**38b.** Coronoid process (ulna)
33e. Coracoid process (scapula)	**92.** Lateral lumbocostal arch
33f. Supraglenoid tubercle (scapula)	**93.** Medial lumbocostal arch

SKELETON—POSTERIOR VIEW

(Enlargement of lumbar vertebrae)

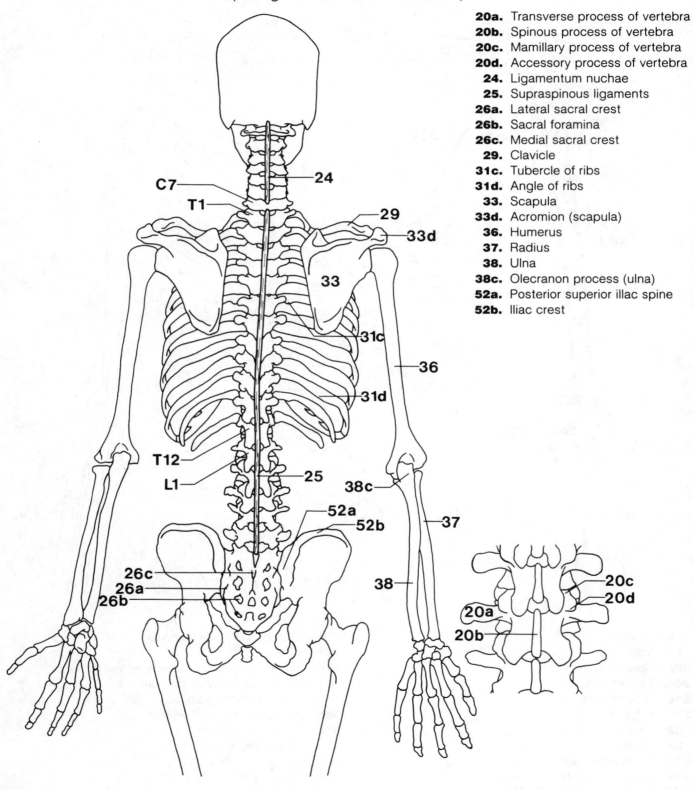

20a. Transverse process of vertebra
20b. Spinous process of vertebra
20c. Mamillary process of vertebra
20d. Accessory process of vertebra
24. Ligamentum nuchae
25. Supraspinous ligaments
26a. Lateral sacral crest
26b. Sacral foramina
26c. Medial sacral crest
29. Clavicle
31c. Tubercle of ribs
31d. Angle of ribs
33. Scapula
33d. Acromion (scapula)
36. Humerus
37. Radius
38. Ulna
38c. Olecranon process (ulna)
52a. Posterior superior iliac spine
52b. Iliac crest

RIGHT ARM—POSTERIOR VIEW

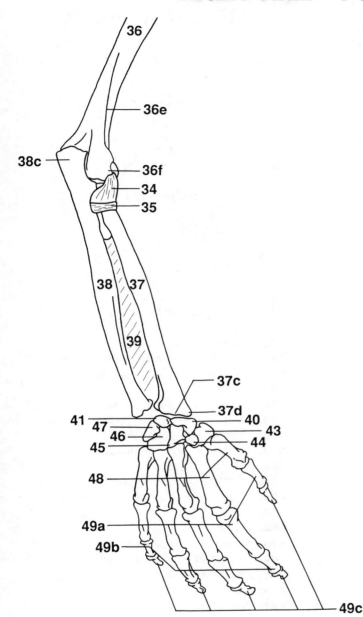

34. Radial collateral ligament
35. Annular ligament
36. Humerus
36e. Lateral supracondylar ridge (humerus)
36f. Lateral epicondyle (humerus)
37. Radius
37c. Dorsal tubercle (radius)
37d. Styloid process (radius)
38. Ulna
38c. Olecranon process (ulna)
39. Interosseous membrane
40. Scaphoid (navicular)
41. Lunate
43. Trapezium
44. Trapezoid
45. Capitate
46. Hamate
47. Triquetrum
48. Metacarpals
49a. Proximal phalanges
49b. Middle phalanges
49c. Distal phalanges

RIGHT ARM
ANTERIOR VIEW

RIGHT HAND
ANTERIOR VIEW

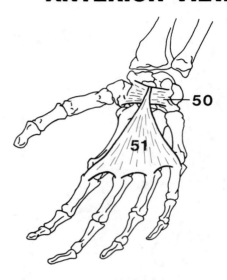

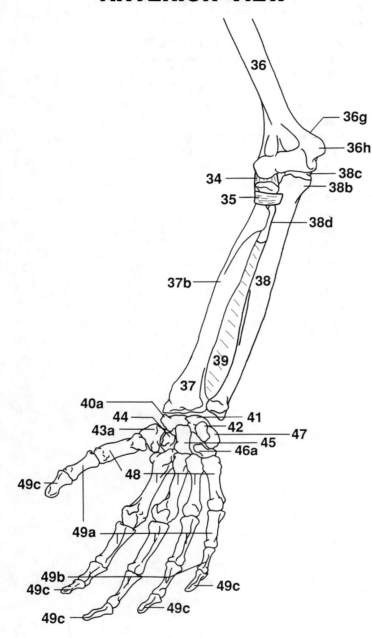

34. Radial collateral ligament
35. Annular ligament
36. Humerus
36g. Medial supracondylar ridge (humerus)
36h. Medial epicondyle (humerus)
37. Radius
37b. Pronator tuberosity (radius)
38. Ulna
38b. Coronoid process (ulna)
38c. Olecranon process (ulna)
38d. Supinator crest (ulna)
39. Interosseous membrane
40a. Tubercle of scaphoid (navicular)
41. Lunate
42. Pisiform
43a. Tubercle of trapezium
44. Trapezoid
45. Capitate
46a. Hook of hamate
47. Triquetrum
48. Metacarpals
49a. Proximal (first) phalanges
49b. Middle (second) phalanges
49c. Distal (third) phalanges
50. Flexor retinaculum
51. Palmar aponeurosis

LUMBAR AND PELVIC REGION—ANTERIOR VIEW

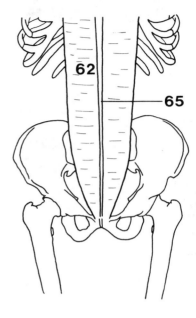

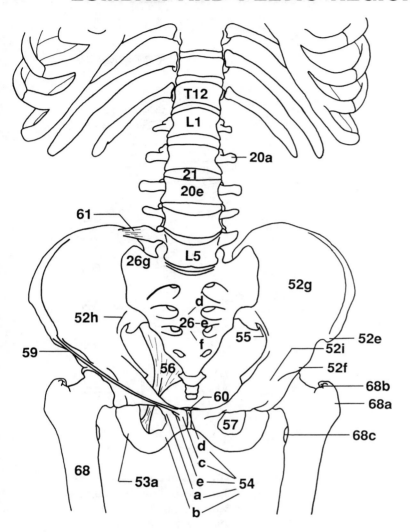

20a. Transverse process of vertebra
20e. Body of vertebra
 21. Intervertebral disk
26d. Second sacral vertebra
26e. Third sacral vertebra
 26f. Fourth sacral vertebra
26g. Ala of sacrum
52e. Anterior superior iliac spine
 52f. Anterior inferior iliac spine
52g. Iliac fossa
52h. Arcuate line (ilium)
 52i. Iliopectineal eminence (ilium)
53a. Ramus of ischium
54a. Superior ramus of pubis
54b. Inferior ramus of pubis
54c. Pubic crest
54d. Pubic symphysis
54e. Pubic tubercle
 55. Greater sciatic foramen
 56. Sacrotuberous ligament
 57. Obturator foramen
 59. Inguinal ligament
 60. Superior pubic ligament
 61. Iliolumbar ligament
 62. Rectus sheath
 65. Linea alba
 68. Femur
68a. Greater trochanter (femur)
68b. Trochanteric fossa (femur)
68c. Lesser trochanter (femur)

PELVIC GIRDLE TO KNEE LATERAL VIEW

52c
52b
52e

68

67

72

74c

THORACIC TO PELVIC REGION LATERAL VIEW

(Arm and leg removed)

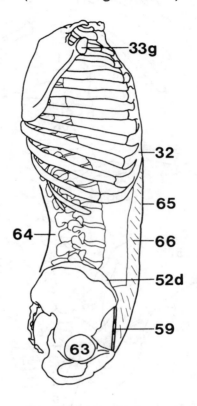

33g

32

65

64

66

52d

59

63

32. Costal cartilage
33g. Glenoid cavity (scapula)
52d. Anterior iliac crest
59. Inguinal ligament
63. Acetabulum
64. Thoracolumbar fascia
65. Linea alba
66. Abdominal aponeurosis

52b. Iliac crest
52c. Iliac tubercle
52e. Anterior superior iliac spine
67. Iliotibial tract
68. Femur
72. Synovial membrane of knee joint
74c. Lateral condyle of tibia

PELVIC GIRDLE—POSTERIOR VIEW

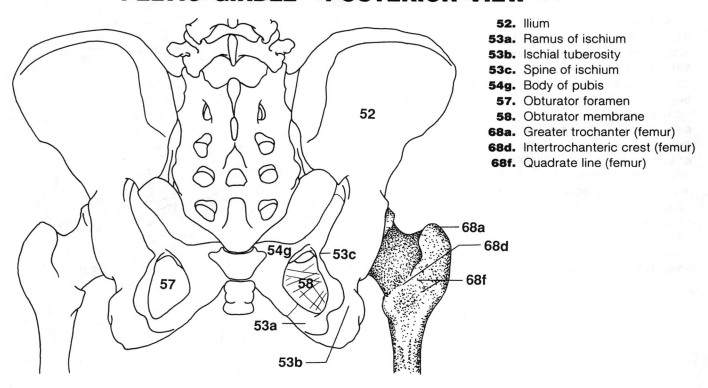

52. Ilium
53a. Ramus of ischium
53b. Ischial tuberosity
53c. Spine of ischium
54g. Body of pubis
57. Obturator foramen
58. Obturator membrane
68a. Greater trochanter (femur)
68d. Intertrochanteric crest (femur)
68f. Quadrate line (femur)

PELVIC GIRDLE AND UPPER LEG— THREE-QUARTER POSTERIOR VIEW

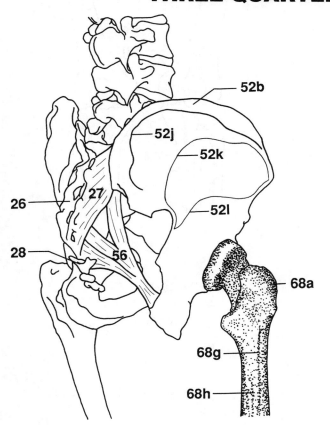

26. Sacrum
27. Aponeurosis of erector spinae
28. Coccyx
52b. Iliac crest
52j. Posterior gluteal line (ilium)
52k. Middle (anterior) gluteal line (ilium)
52l. Inferior gluteal line (ilium)
56. Sacrotuberous ligament
68a. Greater trochanter (femur)
68g. Gluteal tuberosity of femur
68h. Linea aspera (femur)

PELVIC GIRDLE TO LEG—ANTERIOR VIEW

52. Ilium
52e. Anterior superior iliac spine
52f. Anterior inferior iliac spine
54b. Inferior ramus of pubis
54f. Pectineal line (pubis)
54g. Body of pubis
68. Femur
68a. Greater trochanter (femur)
68c. Lesser trochanter (femur)
68e. Intertrochanteric line (femur)
68i. Lateral supracondylar line (femur)
68j. Medial supracondylar line (femur)
69. Quadriceps tendon
70. Patella
71. Patellar ligament
74. Tibia
74a. Tuberosity of tibia

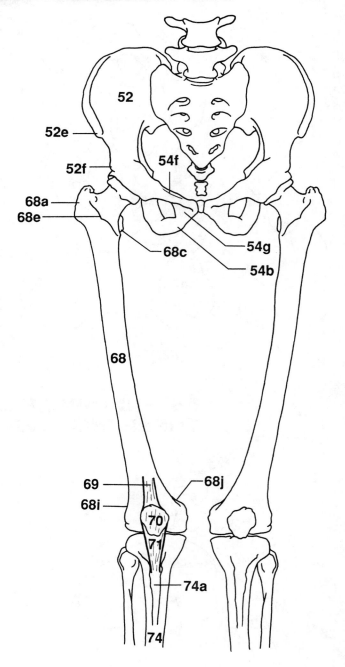

PELVIC GIRDLE TO LEG—POSTERIOR VIEW

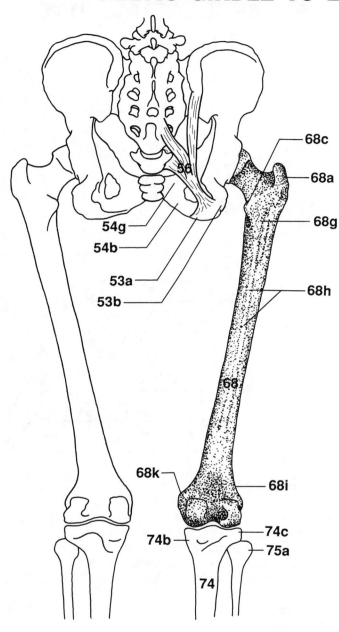

53a. Ramus of ischium
53b. Ischial tuberosity
54b. Inferior ramus of pubis
54g. Body of pubis
56. Sacrotuberous ligament
68. Femur
68a. Greater trochanter (femur)
68c. Lesser trochanter (femur)
68g. Gluteal tuberosity (femur)
68h. Linea aspera (femur)
68i. Lateral supracondylar ridge (line) (femur)
68k. Adductor tubercle (femur)
74. Tibia
74b. Medial condyle of tibia
74c. Lateral condyle of tibia
75a. Head of fibula

RIGHT LEG
ANTEROLATERAL VIEW

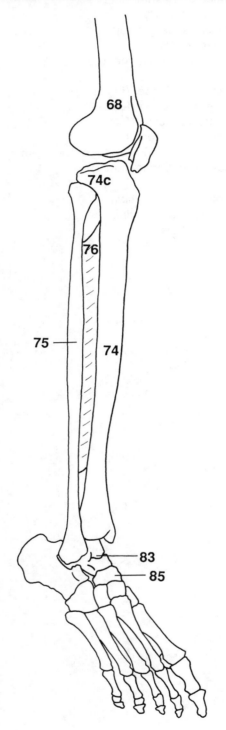

RIGHT FOOT
ANTEROLATERAL VIEW

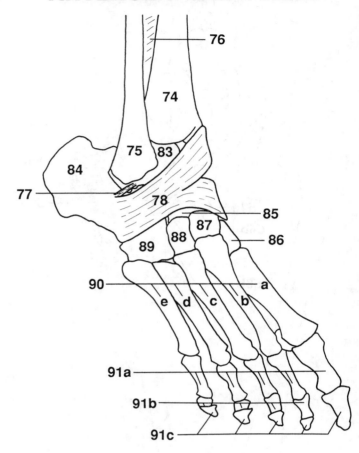

68. Femur
74. Tibia
74c. Lateral condyle of tibia
75. Fibula
76. Interosseous membrane
77. Lateral talocalcaneal ligament
78. Inferior extensor retinaculum
83. Talus
84. Calcaneus
85. Navicular
86. Medial cuneiform
87. Intermediate cuneiform
88. Lateral cuneiform
89. Cuboid
90. Metatarsal bones
90a. First metatarsal
90b. Second metatarsal
90c. Third metatarsal
90d. Fourth metatarsal
90e. Fifth metatarsal
91a. Proximal phalanges
91b. Middle phalanges
91c. Distal phalanges

RIGHT LEG—POSTERIOR VIEW

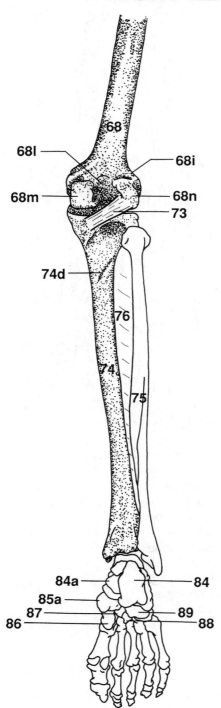

68. Femur
68i. Lateral supracondylar ridge (line) (femur)
68l. Popliteal surface (femur)
68m. Medial condyle (femur)
68n. Lateral condyle (femur)
73. Oblique popliteal ligament
74. Tibia
74d. Soleal line (tibia)
75. Fibula
76. Interosseous membrane
84. Calcaneus
84a. Sustentaculum tali of calcaneus
85a. Tuberosity of navicular
86. Medial cuneiform
87. Intermediate cuneiform
88. Lateral cuneiform
89. Cuboid

RIGHT FOOT—PLANTAR VIEW

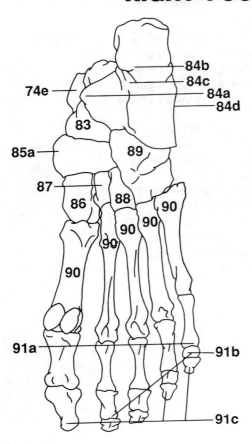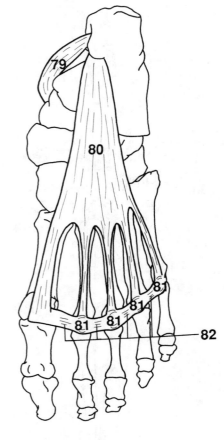

74e. Media malleolus of tibia
79. Flexor retinaculum
80. Plantar aponeurosis
81. Plantar metatarsophalangeal ligaments
82. Transverse metatarsal ligaments
83. Talus
84a. Sustentaculum tali of calcaneus
84b. Tuberosity of calcaneus
84c. Medial border of calcaneus

84d. Lateral border of calcaneus
85a. Tuberosity of navicular
86. Medial cuneiform
87. Intermediate cuneiform
88. Lateral cuneiform
89. Cuboid
90. Metatarsal bones
91a. Proximal phalanges
91b. Middle phalanges
91c. Distal phalanges

CHAPTER TWO
MOVEMENTS OF THE BODY

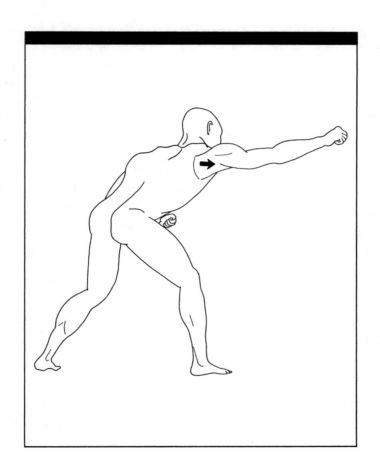

Anatomical position—A subject in the anatomical position is standing erect with the head, eyes, and toes facing forward and the arms hanging straight at the sides with the palms of the hands facing forward.

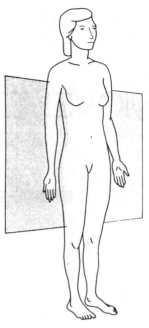

Figure 2.1b
Coronal (frontal) planes—Pass vertically through the body from side to side. They divide the body from front to back.

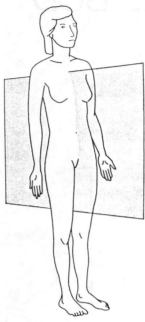

Figure 2.1a
Median or midsagittal plane—Passes vertically through the body from anterior (front) to posterior (back). It divides the body into right and left sides. Other sagittal planes are parallel to this plane.

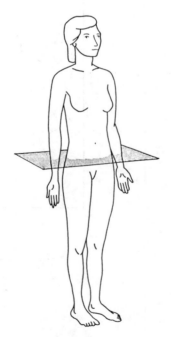

Figure 2.1c
Transverse planes (cross sections)—Pass horizontally through the body parallel to the ground.

Figure 2.2
Flexion—The left arm and forearm, and right thigh are drawn forward in sagittal planes. The right knee is also flexed.
Extension—The left thigh and knee are extended.
Hyperextension—The right arm is hyperextended at the shoulder.

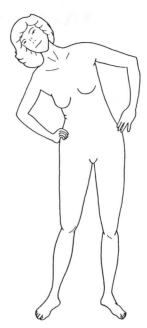

Figure 2.3
Lateral flexion—The torso (or head) bends laterally in the coronal plane.

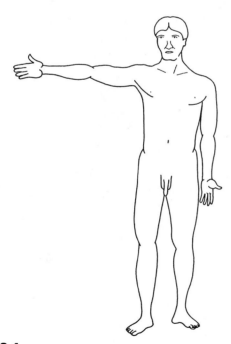

Figure 2.4
Abduction—The right arm is drawn laterally in the coronal plane.
Adduction—The left arm is returned from abduction to the anatomical position.

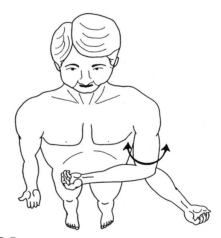

Figure 2.5
Medial rotation—The anterior of the arm (or thigh) is moved toward the median plane.
Lateral rotation—The anterior of the arm (or thigh) is moved away from the median plane.

MOVEMENTS OF THE SCAPULA

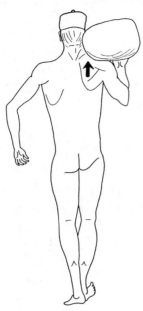

Figure 2.6a
Elevation—The right scapula of this figure is drawn superiorly.

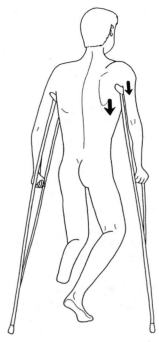

Figure 2.6b
Depression—The right scapula of this figure is pushing the arm inferiorly.

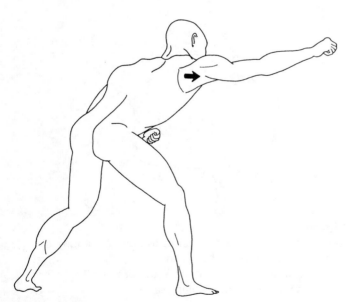

Figure 2.6c
Protraction—The scapula pushes the arm forward in a sagittal plane.

Figure 2.6d
Retraction—The scapula is pulled back from protraction in a sagittal plane. Since the scapula slides around the ribs toward the median plane it becomes adduction.

MOVEMENTS OF THE HAND AND FOREARM

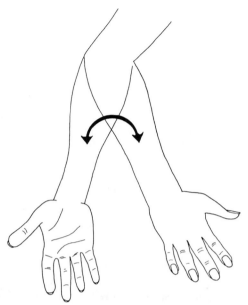

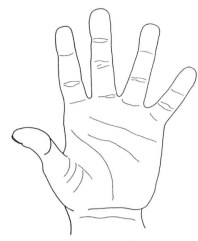

Figure 2.9a

Abduction—The fingers are moved away from the midline of the hand.

Figure 2.7

Pronation—The forearm is rotated away from the anatomical position so that the palm turns medially then posteriorly. If the forearm is flexed at the elbow then the palm turns inferiorly.

Supination—The forearm is rotated so that the palm turns anteriorly (or superiorly if the forearm is flexed).

Figure 2.9b

Adduction—The fingers are moved toward the midline of the hand.

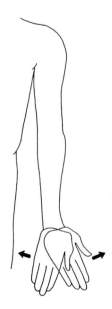

Figure 2.8

Radial flexion (abduction)—The hand, at the wrist, is drawn away from the body in a coronal plane.

Ulnar flexion (adduction)—The hand, at the wrist, is drawn toward the body in a coronal plane.

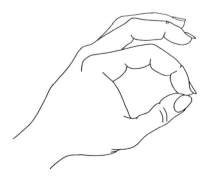

Figure 2.9c

Opposition—The thumb is rotated so its anterior pad can touch the anterior pads of the four fingers.

MOVEMENTS OF THE FOOT

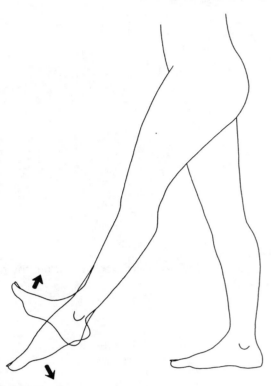

Figure 2.10
Dorsiflexion—The ankle flexes, moving the foot superiorly.
Plantar flexion—The ankle extends, moving the foot
inferiorly.

Figure 2.11a
Eversion—The front of the foot moves laterally away from
the midline (abduction) and the sole turns outward.

Figure 2.11b
Inversion—The front of the foot moves medially toward
the midline (adduction) and the sole turns inward.

CHAPTER THREE
MUSCLES OF THE FACE AND HEAD

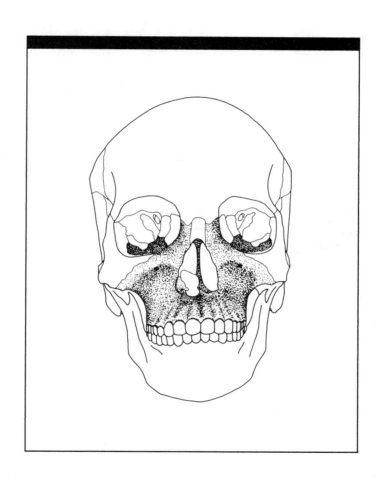

EPICRANIUS

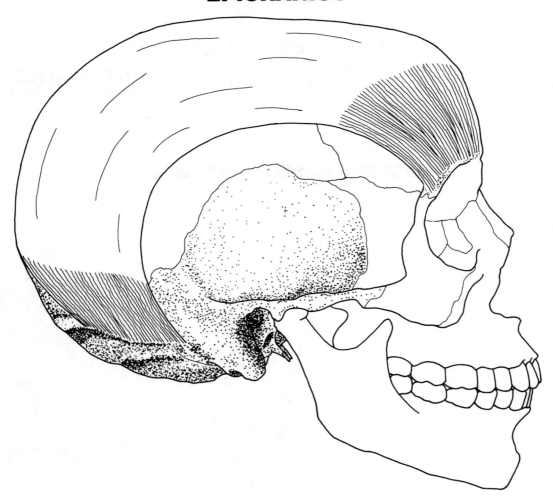

Skull—lateral view

Occipital belly *(occipitalis)*

Origin   Lateral two-thirds of superior nuchal
                  line of occipital bone, mastoid process
                  of temporal bone

Insertion   Galea aponeurotica (an intermediate
                  tendon leading to frontal belly)

Action   Draws back scalp, aids frontal belly to
                  wrinkle forehead and raise eyebrows

Nerve   Posterior auricular branch of facial
                  nerve

Frontal belly *(frontalis)*

Origin   Galea aponeurotica

Insertion   Fascia of facial muscles and skin
                  above nose and eyes

Action   (with occipital belly) Draws back
                  scalp, wrinkles forehead, raises
                  eyebrows

Nerve   Temporal branches of facial nerve

TEMPOROPARIETALIS

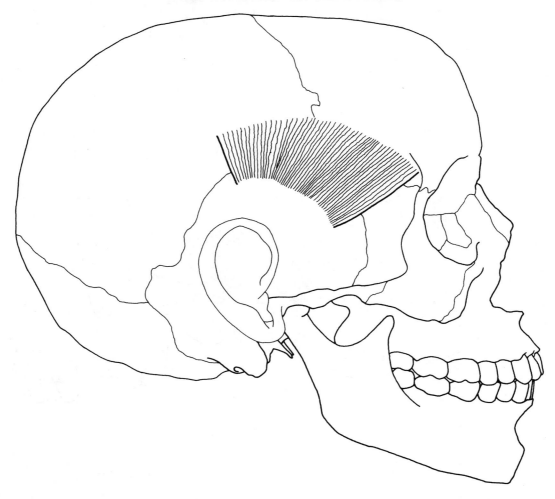

Skull—lateral view

Origin   Fascia over ear   **Action**   Raises ears, tightens scalp
Insertion   Lateral border of galea aponeurotica   **Nerve**   Temporal branch of facial nerve

AURICULARIS ANTERIOR, SUPERIOR, POSTERIOR

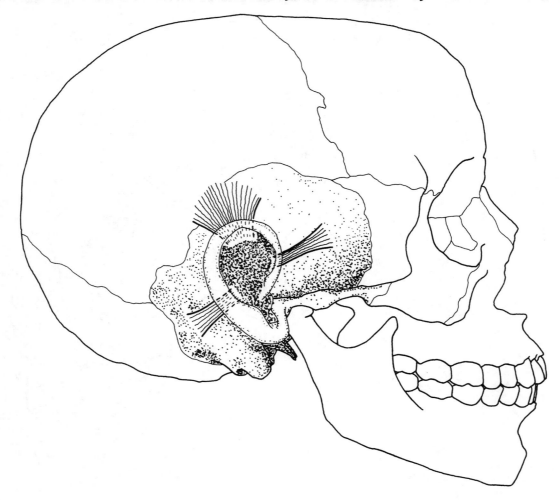

Skull—lateral view

Auricularis anterior

Origin	Fascia in temporal region
Insertion	Anterior to helix of ear
Action	Draws ear forward in some individuals, moves scalp*
Nerve	Temporal branch of facial nerve

Auricularis superior

Origin	Fascia in temporal region
Insertion	Superior part of ear
Action	Draws ear upward in some individuals, moves scalp*
Nerve	Temporal branch of facial nerve

Auricularis posterior

Origin	Mastoid area of temporal bone
Insertion	Posterior part of ear
Action	Draws ear upward in some individuals*
Nerve	Posterior auricular branch of facial nerve

*This muscle is nonfunctional in most people.

ORBICULARIS OCULI

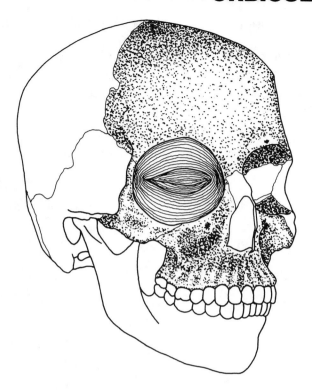

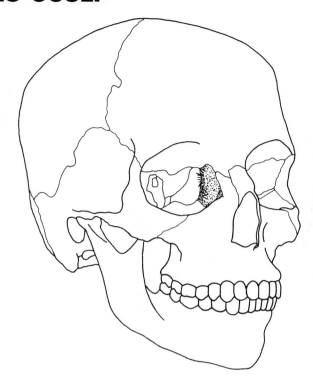

ORBITAL AND PALPEBRAL PARTS

LACRIMAL PART

Skull—three-quarter anterior view

Orbital part

Origin	Frontal bone, maxilla (medial margin of orbit)
Insertion	Continues around orbit and returns to origin
Action	Strong closing of eye
Nerve	Temporal and zygomatic branches of facial nerve

Palpebral part   *(in eyelids)*

Origin	Medial palpebral ligament
Insertion	Lateral palpebral ligament into zygomatic bone
Action	Gentle closure of eyelids
Nerve	Temporal and zygomatic branches of facial nerve

Lacrimal part   *(behind medial palpebral ligament and lacrimal sac)*

Origin	Lacrimal bone
Insertion	Lateral palpebral raphe
Action	Draws lacrimal canals onto surface of eye
Nerve	Temporal and zygomatic branches of facial nerve

LEVATOR PALPEBRAE SUPERIORIS

Origin	Inferior surface of lesser wing of sphenoid
Insertion	Skin of upper eyelid
Action	Raises upper eyelid
Nerve	Oculomotor nerve

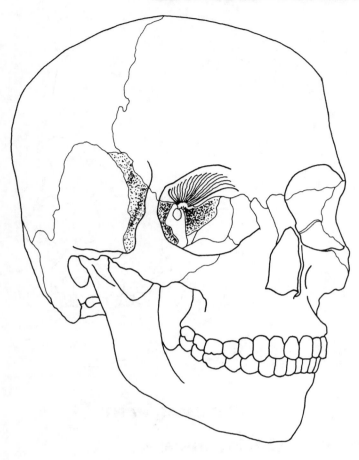

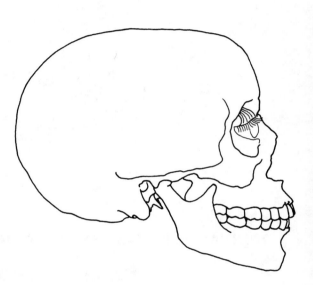

Skull—three-quarter anterior view   **Skull—lateral view**

CORRUGATOR SUPERCILII

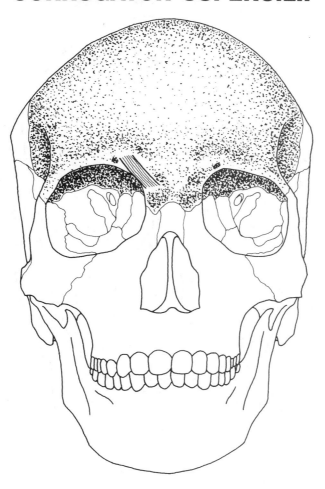

Skull—anterior view

Origin	Medial end of superciliary arch	**Action**	Draws eyebrows downward and medially
Insertion	Deep surface of skin under medial portion of eyebrows	**Nerve**	Temporal branch of facial nerve

PROCERUS

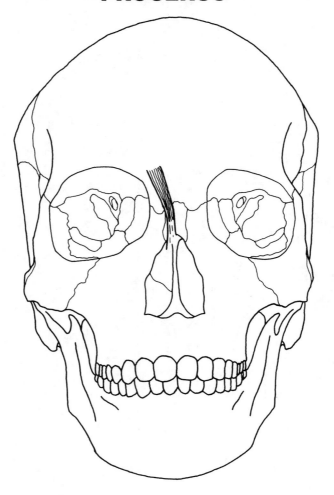

Skull—anterior view

Origin	Fascia over nasal bone and lateral nasal cartilage	**Action**	Draws down medial part of eyebrows, wrinkles nose
Insertion	Skin between eyebrows	**Nerve**	Buccal branches of facial nerve

NASALIS

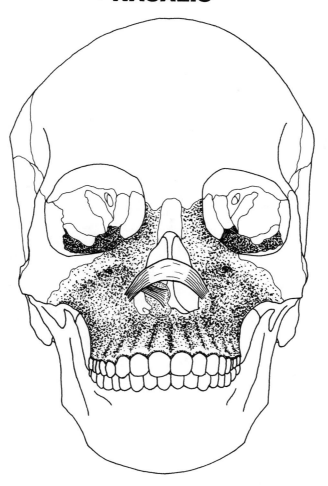

Skull—anterior view

Transverse part

Origin	Middle of maxilla
Insertion	Muscle of opposite side over bridge of nose

Alar part

Origin	Greater alar cartilage, skin on nose
Insertion	Skin at point of nose
Action	Maintains opening of external nares during inspiration
Nerve	Buccal branches of facial nerve

DEPRESSOR SEPTI

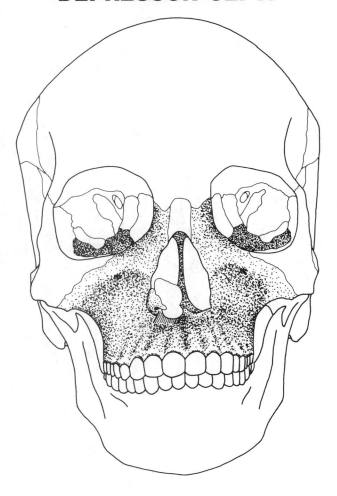

Skull—anterior view

Origin	Incisive fossa of maxilla	**Action**	Constricts nares
Insertion	Nasal septum and ala	**Nerve**	Buccal branches of facial nerve.

ORBICULARIS ORIS

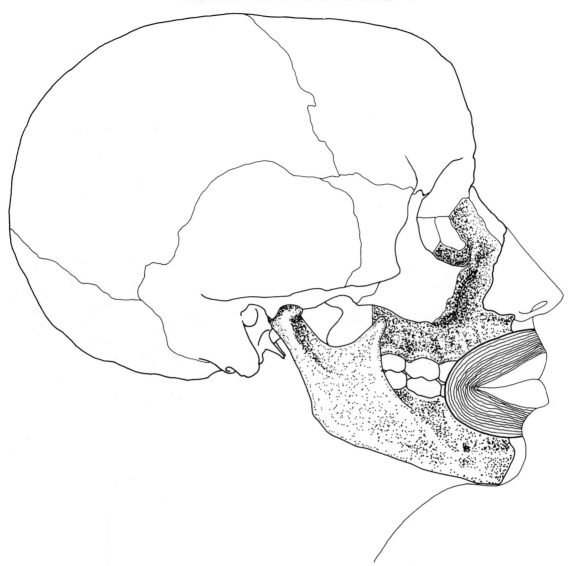

Skull—lateral view

Origin

Lateral band—alveolar border of maxilla
Medial band—septum of nose
Inferior portion—lateral to midline of mandible

Insertion

Becomes continuous with other muscles at angle of mouth

Action

Closure and protrusion of lips

Nerve

Buccal and mandibular branches of facial nerve

LEVATOR LABII SUPERIORIS

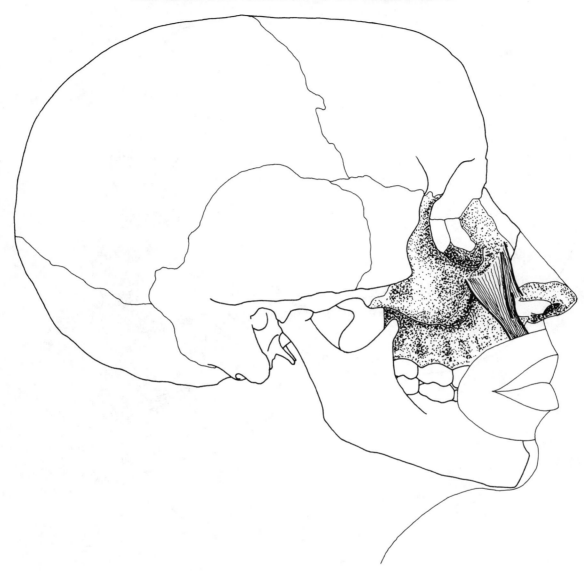

Skull—lateral view

Angular head

Origin   Frontal process of maxilla and zygomatic bone

Insertion Greater alar cartilage and skin of nose, upper lip

Action   Elevates upper lip, dilates nares, forms nasolabial furrow

Nerve   Buccal branches of facial nerve

Infraorbital head

Origin   Lower margin of orbit

Insertion Muscles of upper lip

Action   Elevates upper lip

Nerve   Buccal branches of facial nerve

LEVATOR ANGULI ORIS

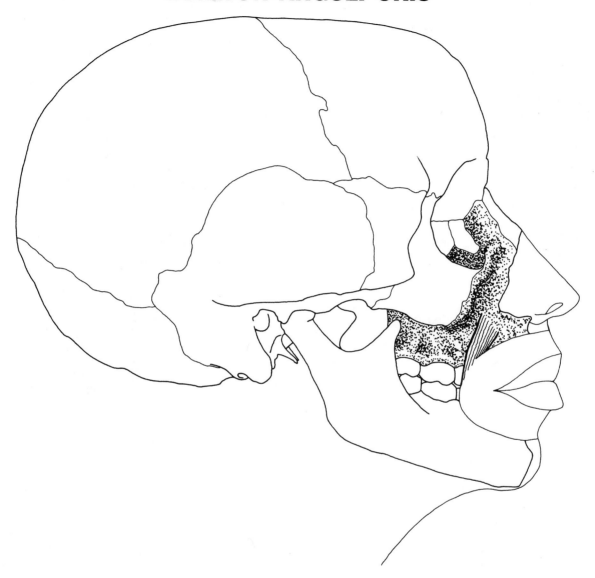

Skull—lateral view

Origin Canine fossa of maxilla
Insertion Angle of mouth

Action Elevates the corner (angle) of the mouth
Nerve Buccal branches of facial nerve

ZYGOMATICUS MAJOR

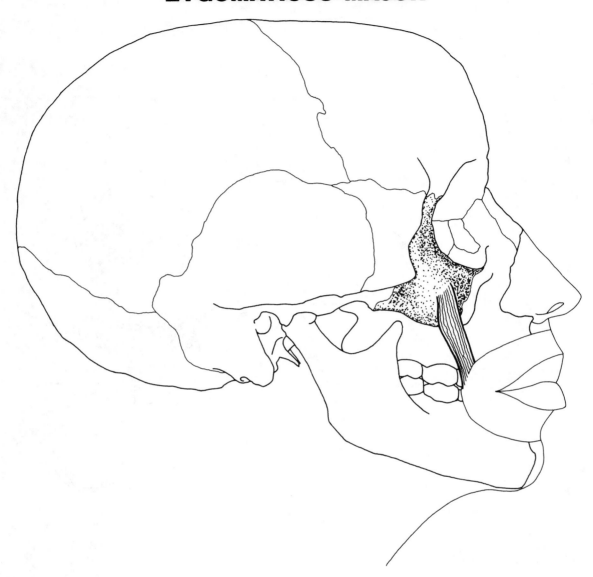

Skull—lateral view

Origin Zygomatic bone

Insertion Angle of mouth

Action Draws angle of mouth upward and backward (laughing)

Nerve Buccal branches of facial nerve

ZYGOMATICUS MINOR

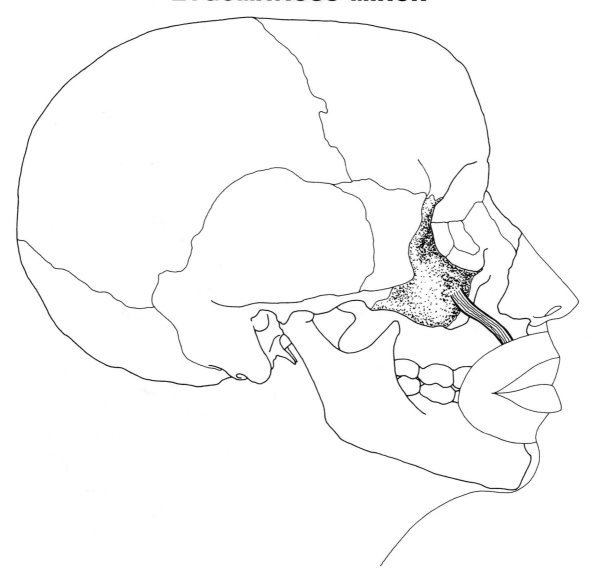

Skull—lateral view

Origin	Zygomatic bone	**Action**	Forms nasolabial furrow
Insertion	Upper lip lateral to levator labii superioris	**Nerve**	Buccal branches of facial nerve

RISORIUS

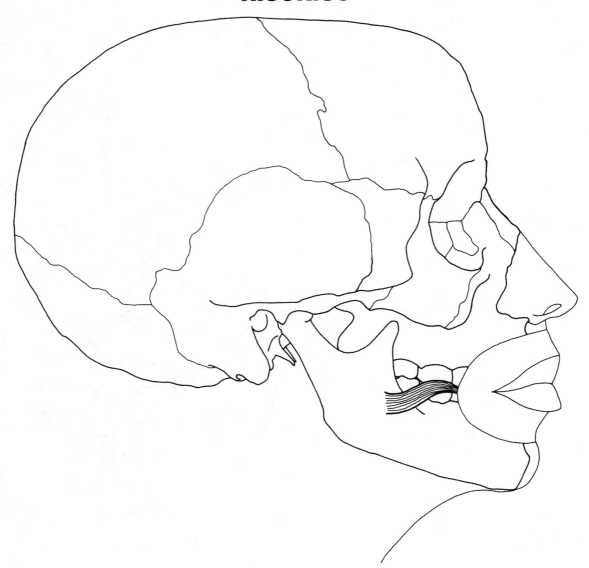

Skull—Lateral View

Origin	Fascia over masseter	**Action**	Retracts angle of mouth, as in grinning
Insertion	Skin at angle of mouth	**Nerve**	Buccal branches of facial nerve

DEPRESSOR LABII INFERIORIS

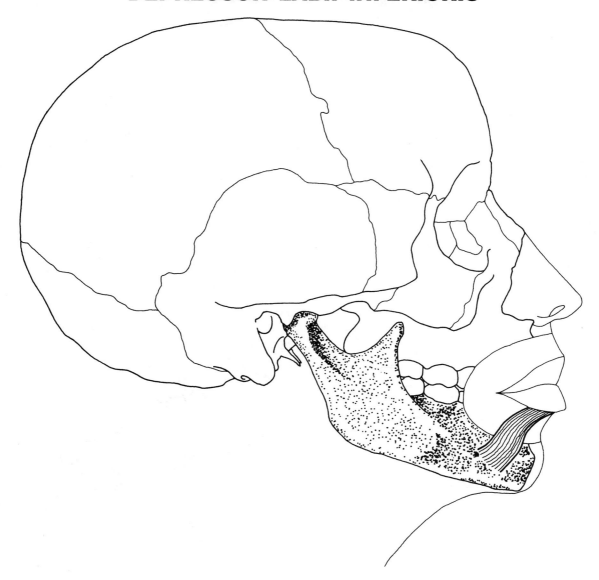

Skull—lateral view

Origin	Mandible, between symphysis and mental foramen	**Action**	Draws lower lip downward and laterally
Insertion	Skin of lower lip	**Nerve**	Mandibular branch of facial nerve

DEPRESSOR ANGULI ORIS

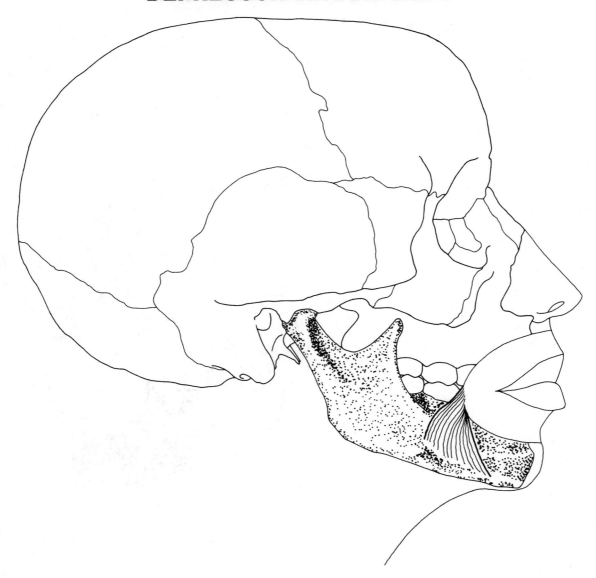

Skull—lateral view

Origin   Oblique line of the mandible

Insertion   Angle of the mouth

Action   Depresses angle of mouth, as in frowning

Nerve   Mandibular branch of facial nerve

MENTALIS

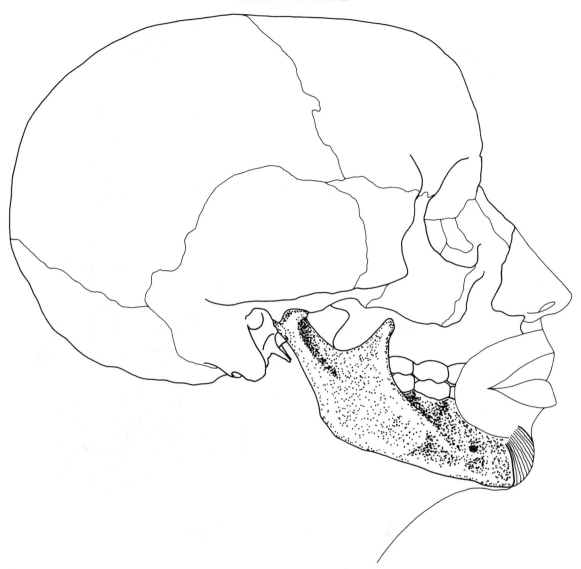

Skull—lateral view

Origin Incisive fossa of mandible

Insertion Skin of chin

Action Raises and protrudes lower lip, wrinkles skin of chin

Nerve Mandibular branch of facial nerve

BUCCINATOR

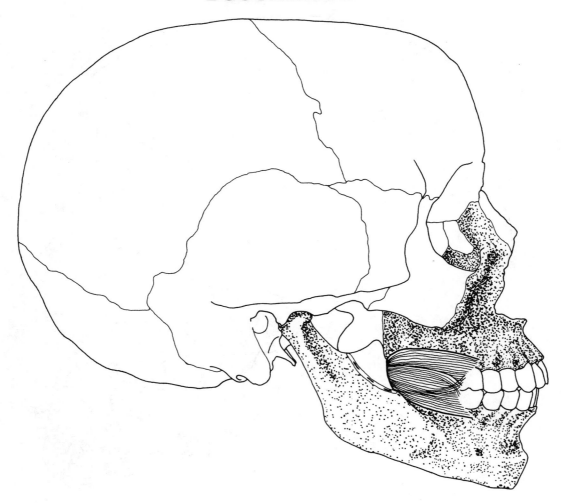

Skull—lateral view

Origin	Outer surface of alveolar processes of maxilla and mandible over molars and along pterygomandibular raphe	**Insertion**	Deep part of muscles of lips
		Action	Compresses cheek
		Nerve	Buccal branches of facial nerve

TEMPORALIS

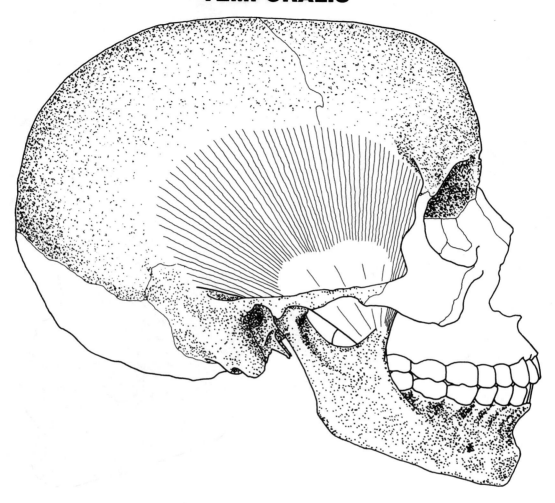

Skull—lateral view

Origin	Temporal fossa including frontal, parietal, and temporal bones	**Action**	Closes lower jaw, clenches teeth
Insertion	Coronoid process and anterior border of ramus of mandible	**Nerve**	Mandibular division of trigeminal nerve

MASSETER

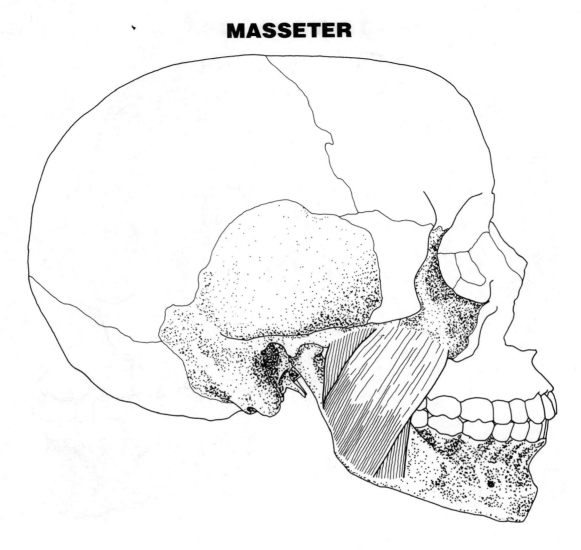

Skull—lateral view

Origin	Zygomatic process of maxilla, medial and inferior surfaces of zygomatic arch	**Action**	Closes lower jaw, clenches teeth
		Nerve	Mandibular division of trigeminal nerve
Insertion	Angle and ramus of mandible, lateral surface of coronoid process of mandible		

PTERYGOIDEUS MEDIALIS

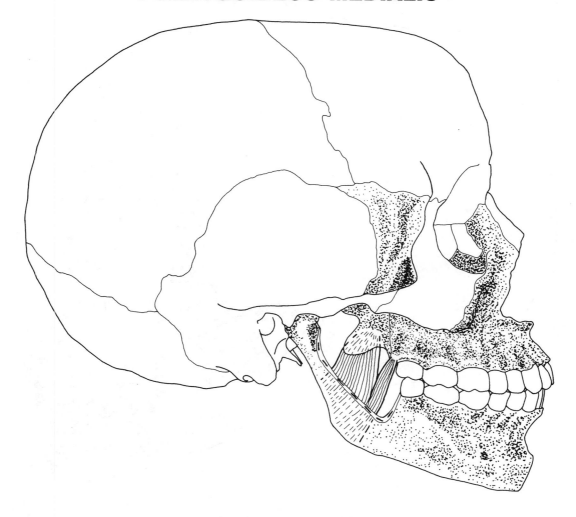

Skull—lateral view

(part of mandible cut away)

Origin   Medial surface of lateral pterygoid plate of sphenoid bone, palatine bone and tuberosity of maxilla

Insertion   Medial surface of ramus and angle of mandible

Action   Closes lower jaw, clenches teeth

Nerve   Mandibular division of trigeminal nerve

PTERYGOIDEUS LATERALIS

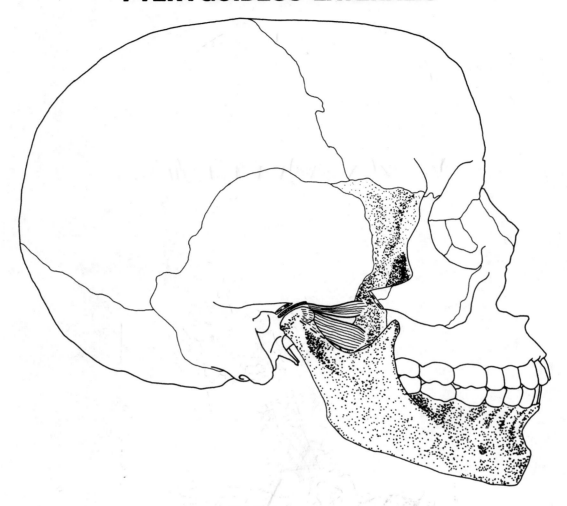

Skull—lateral view

Origin   Superior head—lateral surface of greater wing of sphenoid

Inferior head—lateral surface of lateral pterygoid plate

Insertion   Condyle of mandible, temporomandibular joint

Action   Opens jaws, protrudes mandible, moves mandible sidewards

Nerve   Mandibular division of trigeminal nerve

CHAPTER FOUR
MUSCLES OF THE NECK

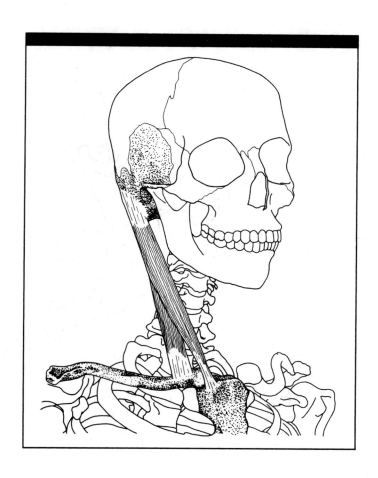

STERNOCLEIDOMASTOIDEUS

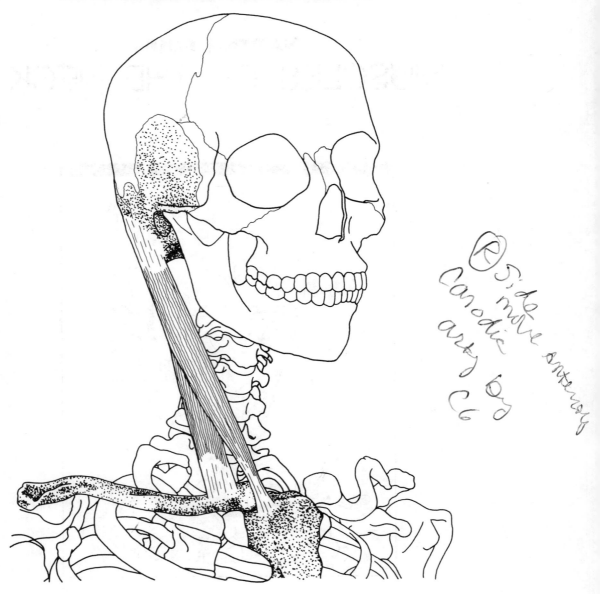

Three-quarter frontal view

Origin	Sternal head—manubrium of sternum Clavicular head—medial part of clavicle	**Action**	One side—bends neck laterally, rotates head Both sides together—flexes neck, draws head ventrally and elevates chin, draws sternum superiorly in deep inspiration
Insertion	Mastoid process of temporal bone, lateral half of superior nuchal line of occipital bone	**Nerve**	Spinal part of accessory nerve (C2,C3)

PLATYSMA

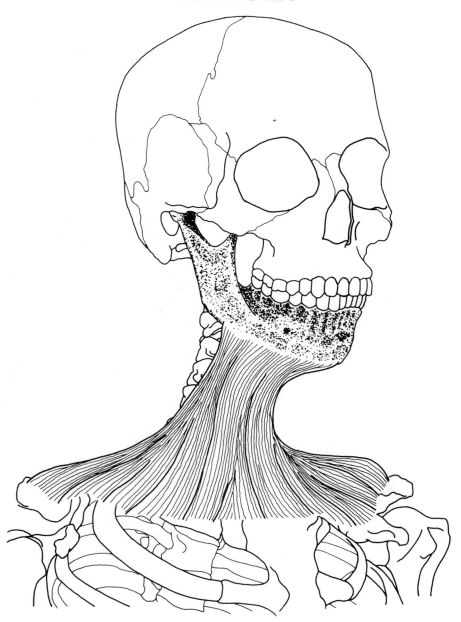

Three-quarter frontal view

Origin   Subcutaneous fascia of upper one-fourth of chest

Insertion   Subcutaneous fascia and muscles of chin and jaw, mandible

Action   Depresses and draws lower lip laterally, draws up skin of chest

Nerve   Cervical branch of facial nerve

DIGASTRICUS

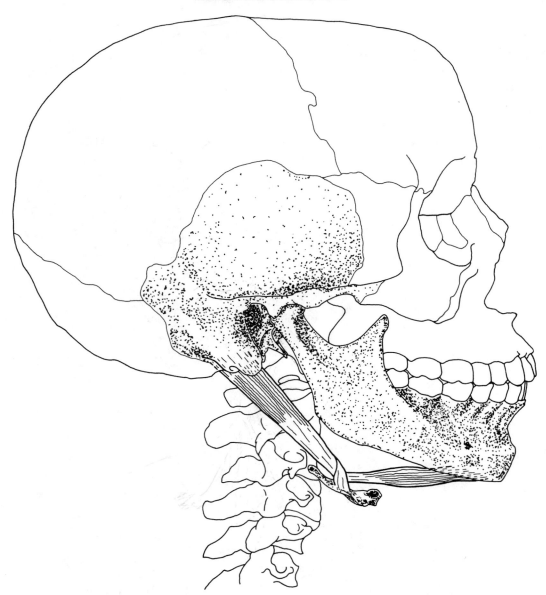

Lateral view

Origin

Posterior belly—mastoid notch of temporal bone
Anterior belly—inner side of inferior border of mandible near symphysis

Insertion

Intermediate tendon attached to hyoid bone

Action

Raises hyoid bone, assists in opening jaws, moves hyoid forward or backward

Nerve

Anterior belly—mandibular division of trigeminal
Posterior belly—facial nerve

STYLOHYOIDEUS

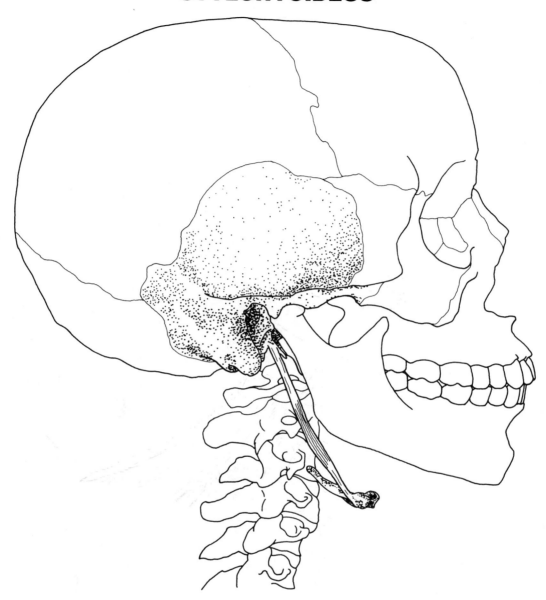

Lateral view

Origin	Styloid process of temporal bone	**Action**	Draws hyoid bone backward, elevates tongue
Insertion	Hyoid bone		
		Nerve	Facial nerve

MYLOHYOIDEUS

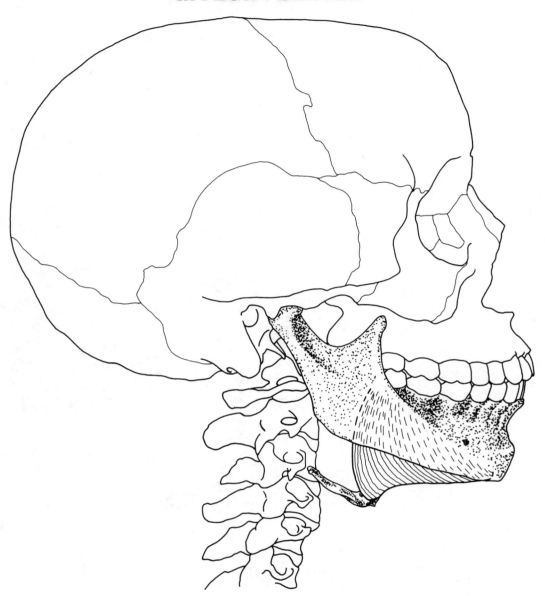

Lateral view

Origin	Inside surface of mandible from symphysis to molars (mylohyoid line)	**Action**	Elevates hyoid bone, raises floor of mouth and tongue
Insertion	Hyoid bone	**Nerve**	Mandibular division of trigeminal nerve

GENIOHYOIDEUS

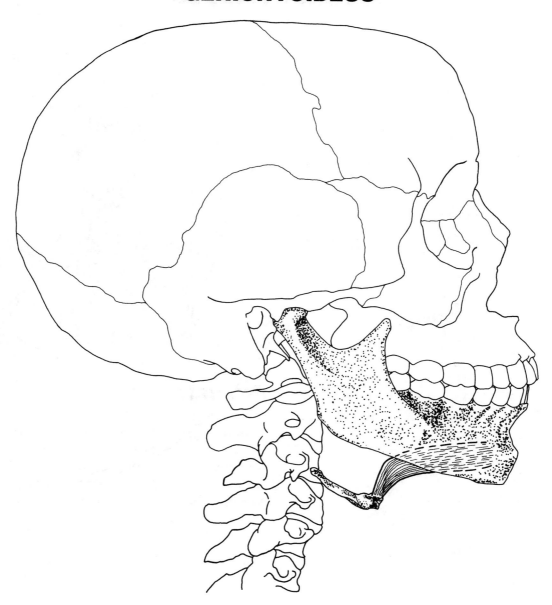

Lateral view

Origin	Inferior mental spine on interior medial surface of mandible	**Action**	Protrudes hyoid bone and tongue
Insertion	Body of hyoid bone	**Nerve**	Branch of C1 through hypoglossal nerve

STERNOHYOIDEUS

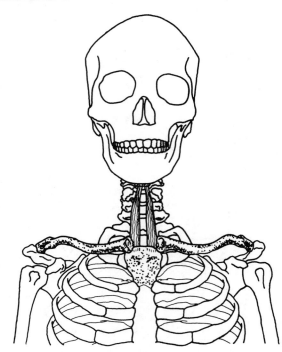

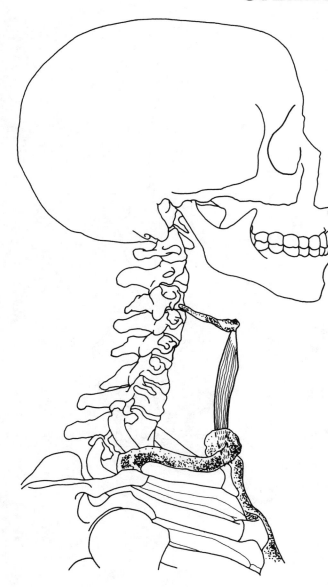

Frontal view

Origin	Medial end of clavicle, manubrium of sternum
Insertion	Body of hyoid bone
Action	Depresses hyoid bone
Nerve	Ansa cervicalis (C1–C3)

Lateral view

STERNOTHYROIDEUS

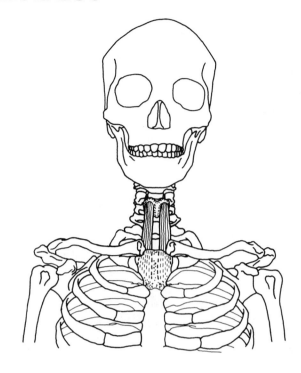

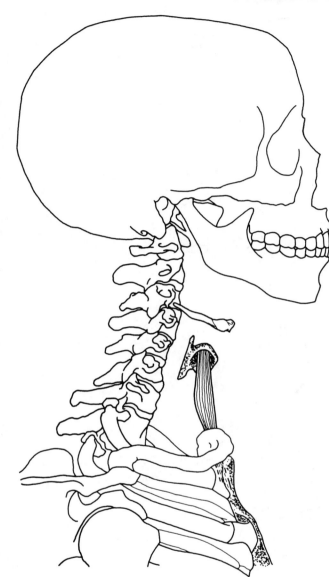

Frontal view

Origin	Dorsal surface of manubrium of sternum
Insertion	Lamina of thyroid cartilage
Action	Depresses thyroid cartilage
Nerve	Ansa cervicalis (C1–C3)

Lateral view

THYROHYOIDEUS

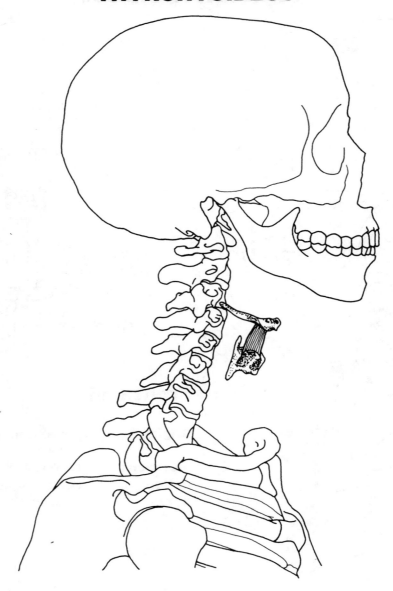

Lateral view

| Origin | Lamina of thyroid cartilage | Action | Depresses hyoid or raises thyroid |
| **Insertion** | Greater cornu of hyoid bone | **Nerve** | C1 through hypoglossal nerve |

OMOHYOIDEUS

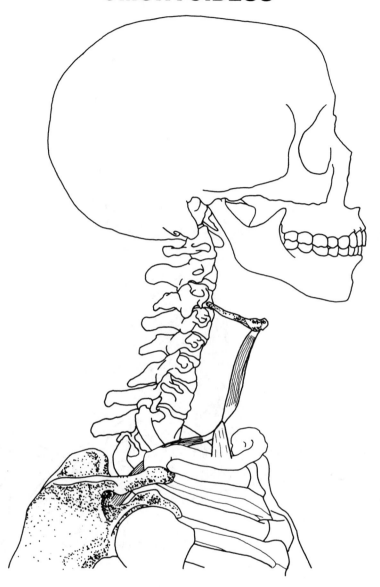

Lateral view

Origin	Superior border of scapula	**Action**	Depresses hyoid bone
Insertion	Inferior belly bound to clavicle by central tendon. Superior belly continues to body of hyoid bone	**Nerve**	Ansa cervicalis (C2,C3)

LONGUS COLLI

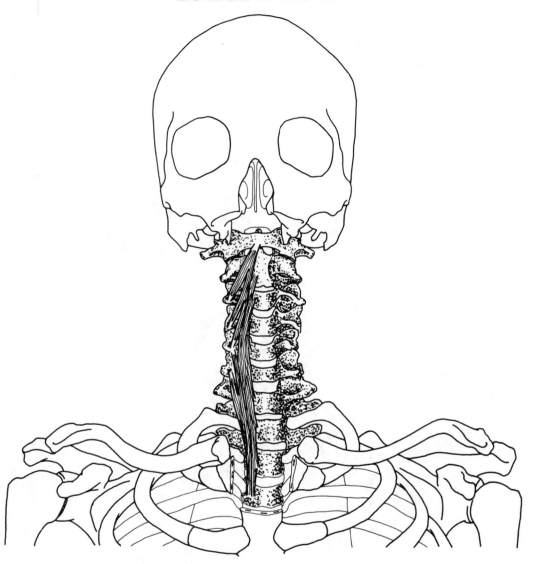

Frontal view

(mandible and part of maxilla removed)

Superior oblique part

Origin	Transverse processes of third, fourth, and fifth cervical vertebrae
Insertion	Anterior arch of atlas

Inferior oblique part

Origin	Anterior surface of bodies of first two or three thoracic vertebrae
Insertion	Transverse processes of fifth and sixth cervical vertebrae

Vertical part

Origin	Anterior surfaces of bodies of upper three thoracic and lower three cervical vertebrae
Insertion	Anterior surfaces of the second, third, and fourth cervical vertebrae
Action	All three parts flex cervical vertebrae
Nerve	C2–C7

LONGUS CAPITIS

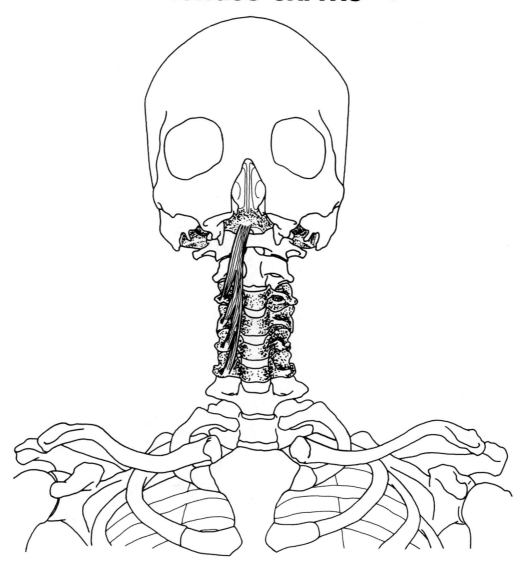

Frontal view

(Mandible and part of maxilla removed)

Origin   Transverse processes of third through sixth cervical vertebrae

Insertion   Occipital bone anterior to foramen magnum

Action   Flexes head

Nerve   C1–C3

RECTUS CAPITIS ANTERIOR

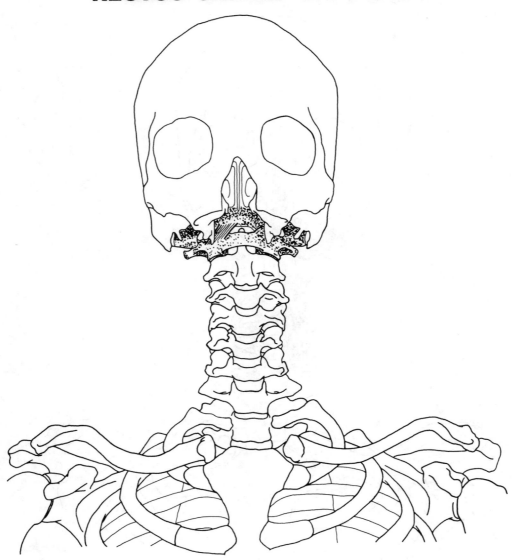

Frontal view

(Mandible and part of maxilla removed)

Origin	Anterior base of transverse process of atlas	**Action**	Flexes head
Insertion	Occipital bone anterior to foramen magnum	**Nerve**	C2,C3

RECTUS CAPITIS LATERALIS

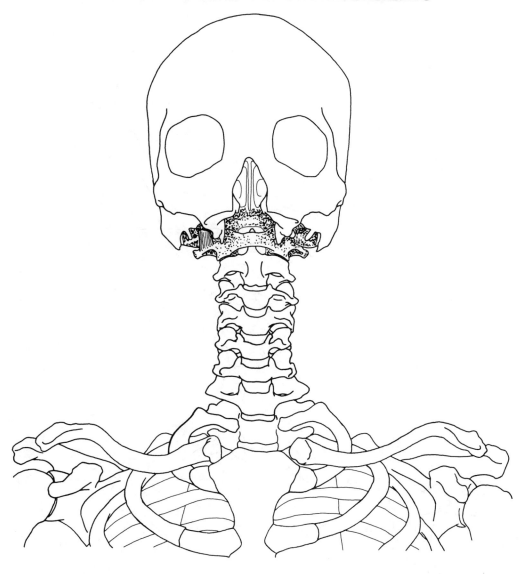

Frontal view

(Mandible and part of maxilla removed)

Origin Transverse process of atlas

Insertion Jugular process of occipital bone

Action Bends head laterally

Nerve C2,C3

SCALENUS ANTERIOR

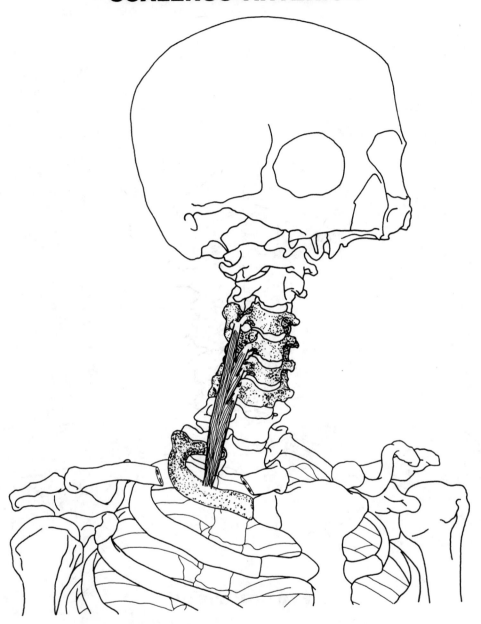

Three-quarter frontal view

(Mandible and part of maxilla removed)

Origin	Transverse processes of third through sixth cervical vertebrae	**Action**	Raises first rib (inspiration), flexes and rotates neck
Insertion	Inner border of first rib (scalene tubercle)	**Nerve**	Ventral rami of cervical nerves

SCALENUS MEDIUS

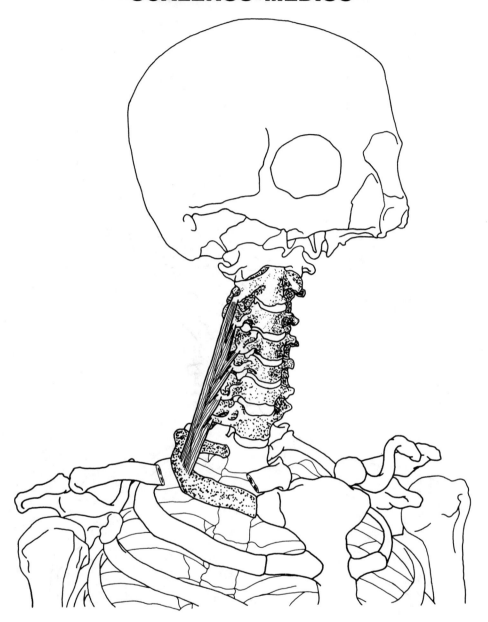

Three-quarter frontal view
(Mandible and part of maxilla removed)

| **Origin** | Transverse processes of lower six cervical vertebrae | **Action** | Raises first rib (inspiration), flexes and rotates neck |
| **Insertion** | Upper surface of first rib | **Nerve** | Ventral rami of cervical nerves |

SCALENUS POSTERIOR

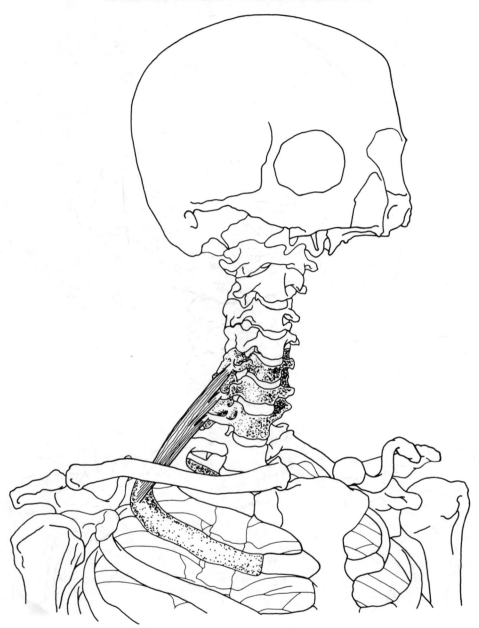

Three-quarter frontal view
(Mandible and part of maxilla removed)

Origin	Transverse processes of lower two or three cervical vertebrae	**Action**	Raises second rib (inspiration), flexes and rotates neck
Insertion	Outer surface of second rib	**Nerve**	Ventral rami of lower cervical nerves

RECTUS CAPITIS POSTERIOR MAJOR

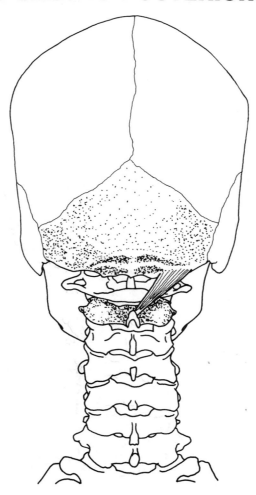

Posterior skull and cervical vertebrae

Origin	Spinous process of axis	**Action**	Extends and rotates head
Insertion	Lateral portion of inferior nuchal line of occipital bone	**Nerve**	Suboccipital nerve

RECTUS CAPITIS POSTERIOR MINOR

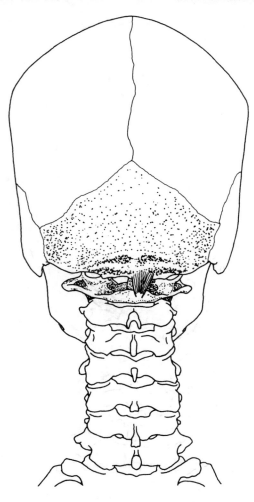

Posterior skull and cervical vertebrae

Origin	Posterior arch of atlas	**Action**	Extends head
Insertion	Medial portion of inferior nuchal line of occipital bone	**Nerve**	Suboccipital nerve

OBLIQUUS CAPITIS INFERIOR

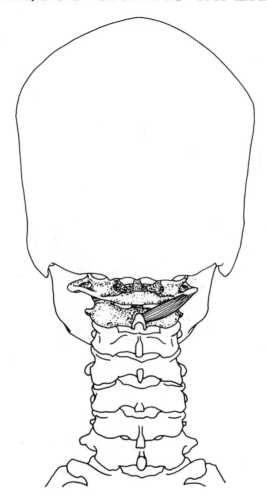

Posterior skull and cervical vertebrae

Origin	Spinous process of axis	**Action**	Rotates atlas
Insertion	Transverse process of atlas	**Nerve**	Suboccipital nerve

OBLIQUUS CAPITIS SUPERIOR

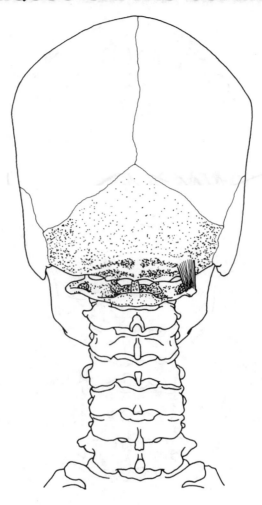

Posterior skull and cervical vertebrae

Origin   Transverse process of atlas

Insertion   Occipital bone between inferior and superior nuchal lines

Action   Extends and bends head laterally

Nerve   Suboccipital nerve

CHAPTER FIVE
MUSCLES OF THE TRUNK

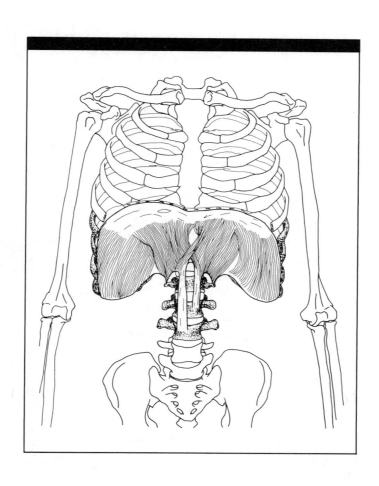

SPLENIUS CAPITIS

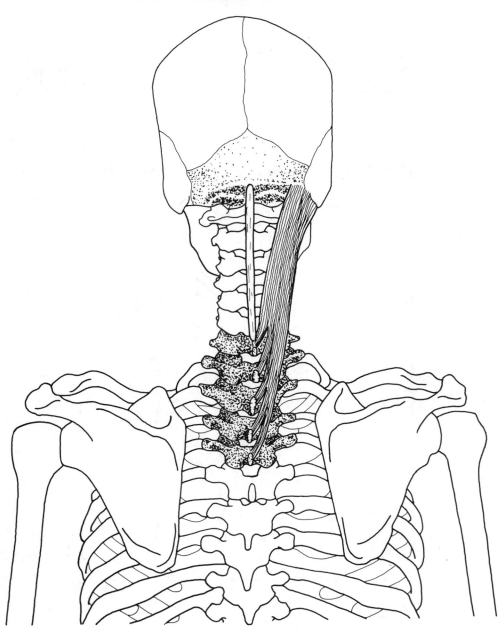

Posterior skull, neck, and back

Origin   Lower part of ligamentum nuchae,
spinous processes of seventh cervical
vertebra, and upper three or four
thoracic vertebrae

Insertion   Mastoid process of temporal bone
and lateral part of superior nuchal line

Action   Extends and rotates the head

Nerve   Lateral branches of dorsal primary
divisions of middle and lower cervical
nerves

SPLENIUS CERVICIS*

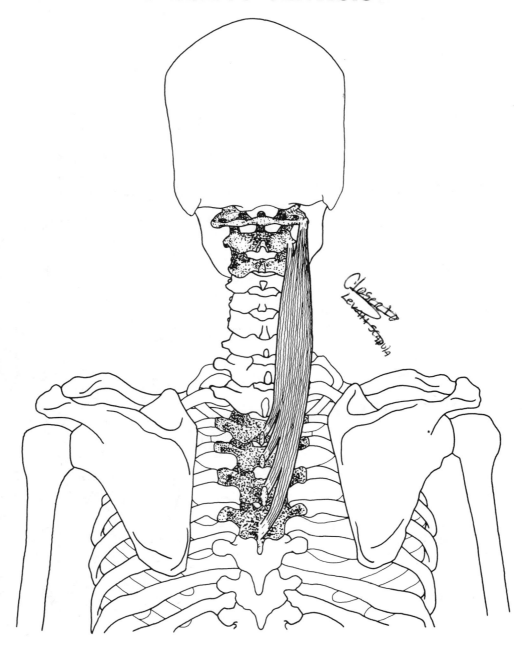

Posterior skull, neck, and back

Origin	Spinous processes of third to sixth thoracic vertebrae	**Action**	Extends and rotates the head
Insertion	Transverse processes of upper two or three cervical vertebrae	**Nerve**	Lateral branches of dorsal primary divisions of middle and lower cervical nerves

* Wraps around longissimus

ILIOCOSTALIS LUMBORUM*

Origin   Medial and lateral sacral crests and medial part of iliac crests

Insertion   Angles of lower six ribs

Action   Extension, lateral flexion of vertebral column, rotates ribs for forceful inspiration

Nerve   Dorsal primary divisions of spinal nerves

ILIOCOSTALIS THORACIS*

Origin   Angles of lower six ribs medial to iliocostalis lumborum

Insertion   Angles of upper six ribs and transverse process of seventh cervical vertebra

Action   Extension, lateral flexion of vertebral column, rotates ribs for forceful inspiration

Nerve   Dorsal primary divisions of spinal nerves

ILIOCOSTALIS CERVICIS*

Origin   Angles of third through sixth ribs

Insertion   Transverse processes of fourth, fifth, and sixth cervical vertebrae

Action   Extension, lateral flexion of vertebral column

Nerve   Dorsal primary divisions of spinal nerves

* Part of erector spinae along with spinalis and longissimus muscles. The complete origin of erector spinae is: medial and lateral sacral crests and medial part of iliac crests, spinous processes and supraspinal ligament of lumbar, and eleventh and twelfth thoracic vertebrae.

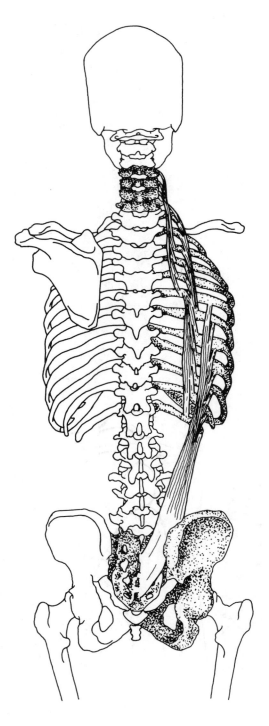

Trunk—dorsal view

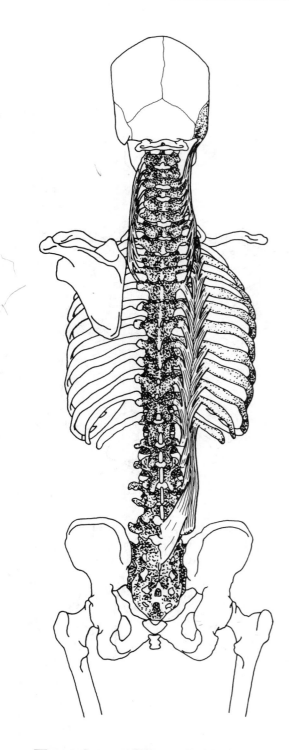

LONGISSIMUS THORACIS*

Origin	Medial and lateral sacral crests, spinous processes and supraspinal ligament of lumbar and eleventh and twelfth thoracic vertebrae, and medial part of iliac crests
Insertion	Transverse processes of all thoracic vertebrae, between tubercles and angles of lower nine or ten ribs
Action	Extension, lateral flexion of vertebral column, rotates ribs for forceful inspiration
Nerve	Dorsal primary divisions of spinal nerves

LONGISSIMUS CERVICIS*

Origin	Transverse processes of upper four or five thoracic vertebrae
Insertion	Transverse processes of second to sixth cervical vertebrae
Action	Extension, lateral flexion of vertebral column
Nerve	Dorsal primary divisions of spinal nerves

LONGISSIMUS CAPITIS*

Origin	Transverse processes of upper five thoracic vertebrae, articular processes of lower three cervical vertebrae
Insertion	Posterior part of mastoid process of temporal bone
Action	Extends and rotates head
Nerve	Dorsal primary divisions of middle and lower cervical nerves

* Part of erector spinae

Trunk—dorsal view

SPINALIS THORACIS*

Origin	Spinous processes of lower two thoracic and upper two lumbar vertebrae
Insertion	Spinous processes of upper thoracic vertebrae
Action	Extends vertebral column
Nerve	Dorsal primary divisions of spinal nerves

SPINALIS CERVICIS*

Origin	Ligamentum nuchae, spinous process of seventh cervical vertebra
Insertion	Spinous process of axis
Action	Extends vertebral column
Nerve	Dorsal primary divisions of spinal nerves

SPINALIS CAPITIS

*(Medial part of semispinalis capitis)**

* Part of erector spinae

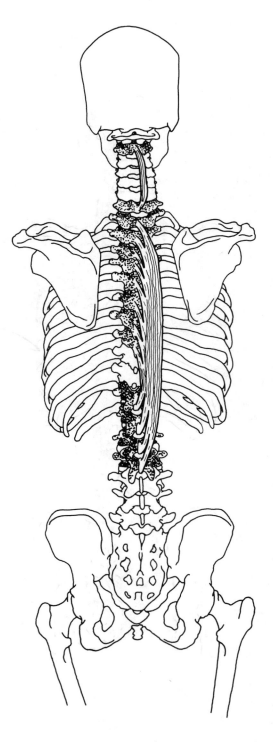

Trunk—dorsal view

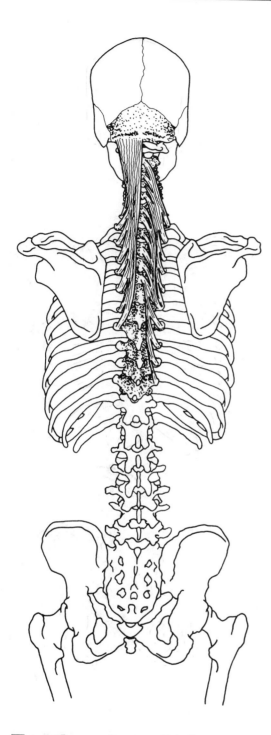

Trunk—dorsal view

SEMISPINALIS THORACIS

Origin	Transverse processes of the sixth through tenth thoracic vertebrae
Insertion	Spinous processes of the lower two cervical and upper four thoracic vertebrae
Action	Extends and rotates vertebral column
Nerve	Dorsal primary divisions of spinal nerves

SEMISPINALIS CERVICIS

Origin	Transverse processes of upper five or six thoracic vertebrae
Insertion	Spinous processes of second to fifth cervical vertebrae
Action	Extends and rotates vertebral column
Nerve	Dorsal primary divisions of spinal nerves

SEMISPINALIS CAPITIS
(Medial part is spinalis capitis)

Origin	Transverse processes of lower four cervical and upper six or seven thoracic vertebrae
Insertion	Between superior and inferior nuchal lines of occipital bone
Action	Extends and rotates head
Nerve	Dorsal primary divisions of spinal nerves

MULTIFIDUS

Origin	Sacral region—along sacral foramina up to posterior, superior iliac spine Lumbar region—mamillary processes of vertebrae Thoracic region—transverse processes Cervical region—articular processes of lower four vertebrae
Insertion	Spinous process two–four vertebrae superior to origin
Action	Extends and rotates vertebral column
Nerve	Dorsal primary division of spinal nerves

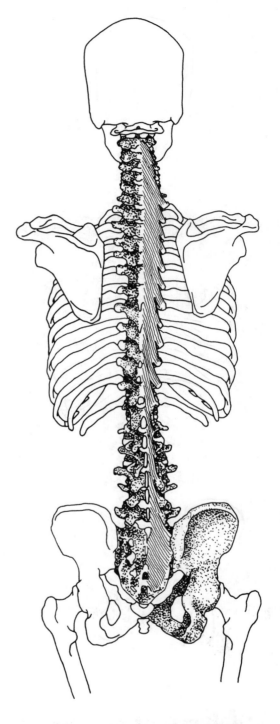

Trunk—dorsal view

ROTATORES

Origin	Transverse process of each vertebra
Insertion	Base of spinous process of next vertebra above
Action	Extends and rotates vertebral column
Nerve	Dorsal primary division of spinal nerves

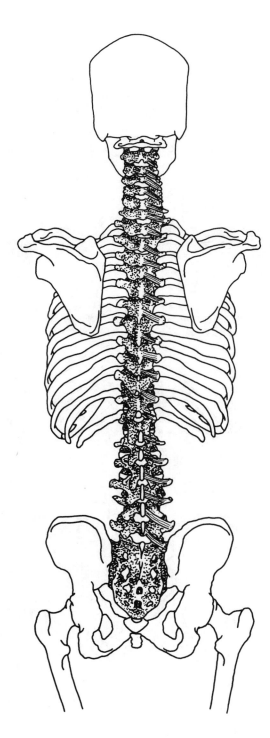

Trunk—dorsal view

INTERSPINALES

(Paired on either side of interspinal ligament)

Origin   Cervical region—spinous processes of third to seventh cervical vertebrae
Thoracic region—spinous processes of second to twelfth vertebrae
Lumbar region—spinous processes of second to fifth lumbar vertebrae

Insertion   Spinous process of next vertebra superior to origin

Action   Extends vertebral column

Nerve   Dorsal primary division of spinal nerves

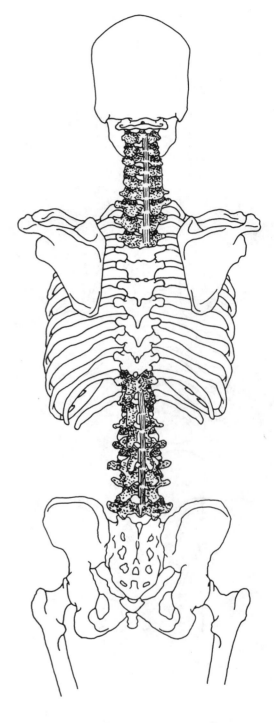

Trunk—dorsal view

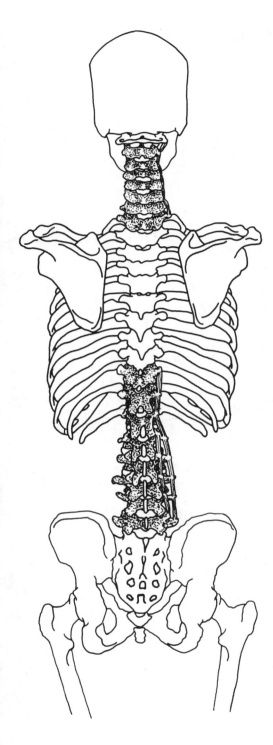

INTERTRANSVERSARII
Cervical region

INTERTRANSVERSARII ANTERIORES

Origin	Anterior tubercle of transverse processes of vertebrae from first thoracic to axis
Insertion	Anterior tubercle of next superior vertebra

INTERTRANSVERSARII POSTERIORES

Origin	Posterior tubercle of transverse processes of vertebrae from first thoracic to axis
Insertion	Posterior tubercle of next superior vertebra

Thoracic region

Origin	Transverse processes of first lumbar to eleventh thoracic vertebrae
Insertion	Transverse processes of next superior vertebra
Action	Lateral flexion of vertebral column
Nerve	Ventral primary division of spinal nerves

Lumbar region

INTERTRANSVERSARII LATERALES

Origin	Transverse processes of lumbar vertebrae
Insertion	Transverse process of next superior vertebra
Action	Lateral flexion of vertebral column
Nerve	Ventral primary division of spinal nerves

INTERTRANSVERSARII MEDIALES

Origin	Mamillary process of each lumbar vertebra
Insertion	Accessory process of the next superior lumbar vertebra
Action	Lateral flexion of vertebral column
Nerve	Dorsal primary division of spinal nerves

Trunk—dorsal view

INTERCOSTALES EXTERNI

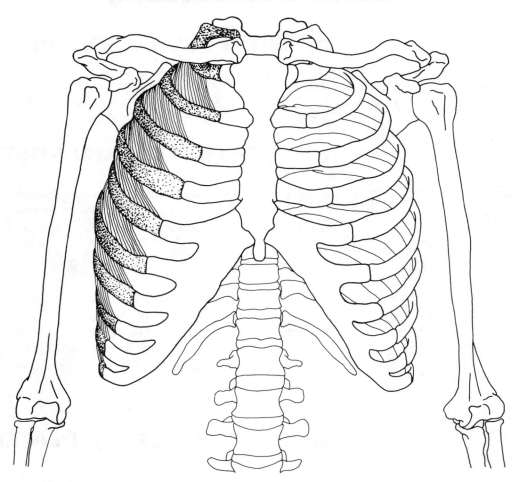

Trunk—anterior view

Origin Lower margin of upper eleven ribs

Insertion Superior border of rib below (each muscle fiber runs obliquely and inserts toward the costal cartilage)

Action Draws ventral part of ribs upward, increasing the volume of the thoracic cavity

Nerve Intercostal nerves

INTERCOSTALES INTERNI

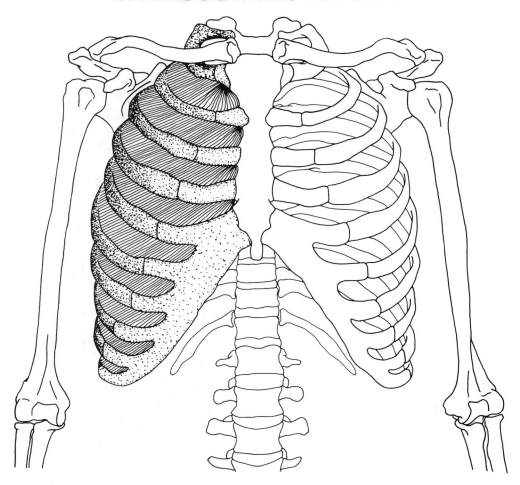

Trunk—anterior view

Origin

From the cartilages to the angles of the upper eleven ribs

Insertion

Superior border of the rib below (each muscle fiber runs obliquely and inserts away from the costal cartilage)

Action

Draws the ventral part of the ribs downward, decreasing the volume of the thoracic cavity

Nerve

Intercostal nerves

SUBCOSTALES

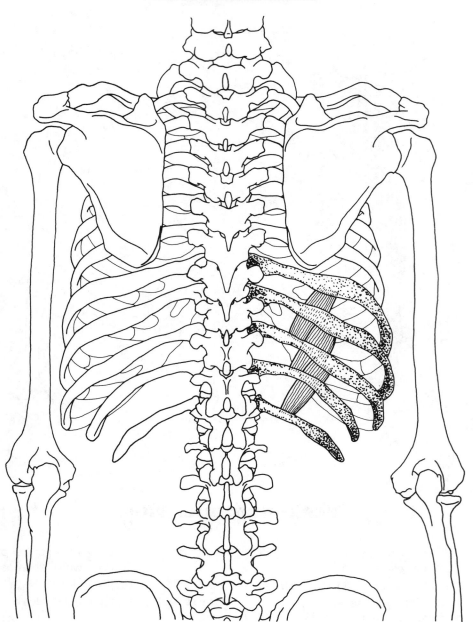

Trunk—dorsal view

Origin	Inner surface of each rib near its angle	**Action**	Draws the ventral part of the ribs downward, decreasing the volume of the thoracic cavity
Insertion	Medially on the inner surface of second or third rib below		
		Nerve	Intercostal nerves

TRANSVERSUS THORACIS

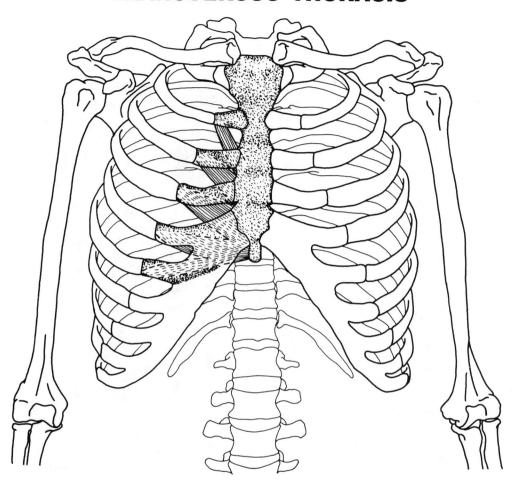

Trunk—anterior view

Origin Inner surface of lower portion of sternum and adjacent costal cartilages

Insertion Inner surfaces of costal cartilages of the second through sixth ribs

Action Draws down the ventral part of the ribs, decreasing the volume of the thoracic cavity

Nerve Intercostal nerves

LEVATORES COSTARUM

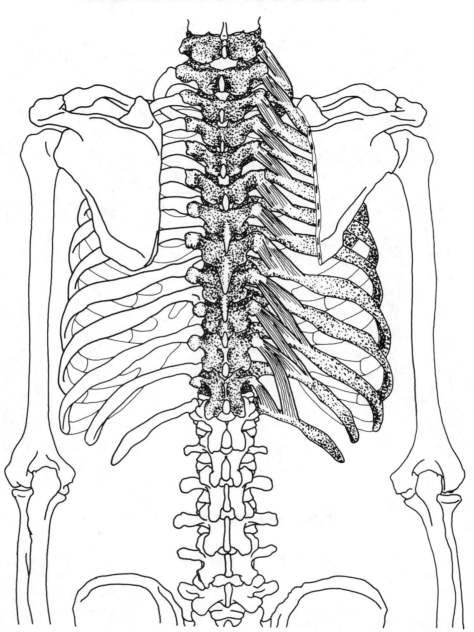

Trunk—dorsal view

Origin	Transverse processes of the seventh cervical and the upper eleven thoracic vertebrae	**Action**	Raises ribs, extends, laterally flexes and rotates vertebral column
		Nerve	Intercostal nerves
Insertion	Laterally to outer surface of next lower rib (lower muscles may cross over one rib)		

SERRATUS POSTERIOR SUPERIOR

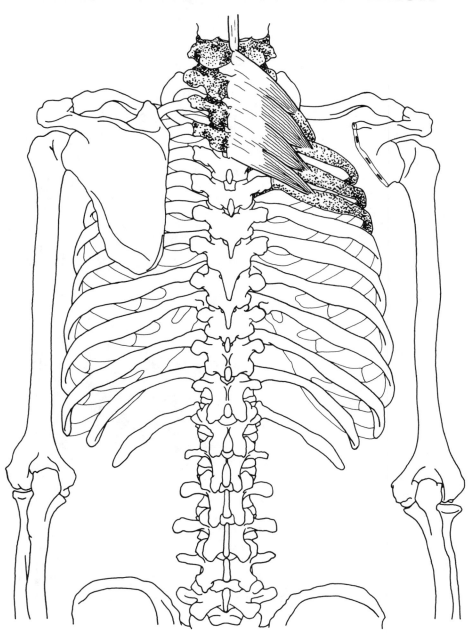

Trunk—dorsal view

Origin	Ligamentum nuchae, spinous processes of seventh cervical and first few thoracic vertebrae	**Action**	Raises ribs in inspiration
		Nerve	T1–T4
Insertion	Upper borders of the second through fifth ribs lateral to their angles		

SERRATUS POSTERIOR INFERIOR

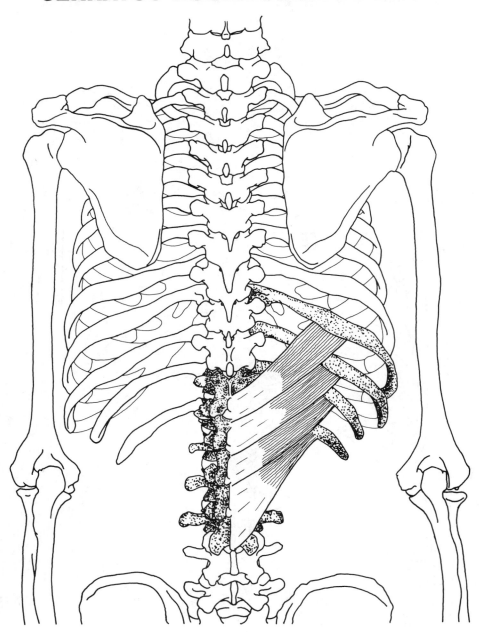

Trunk—dorsal view

Origin	Spinous processes of the lower two thoracic and the upper two or three lumbar vertebrae	**Action**	Pulls ribs down, resisting pull of diaphragm
Insertion	Lower borders of bottom four ribs	**Nerve**	T9–T12

DIAPHRAGM

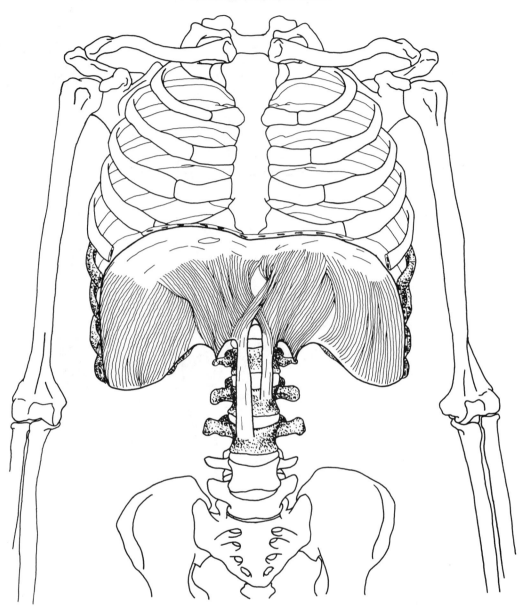

Trunk—anterior view

(Lower costal cartilages removed)

Origin	Sternal part—inner part of xiphoid process Costal part—inner surfaces of lower six ribs and their cartilages Lumbar part—upper two or three lumbar vertebrae and lateral and medial lumbocostal arches	**Insertion**	Central tendon
		Action	Draws central tendon down, increasing volume of thoracic cavity
		Nerve	Phrenic nerve (C3–C5)

OBLIQUUS EXTERNUS ABDOMINIS

Trunk—lateral view

Origin	Lower eight ribs	**Action**	Compresses abdominal contents, laterally flexes and rotates vertebral column
Insertion	Anterior part of iliac crest, abdominal aponeurosis to linea alba		
		Nerve	Eighth to twelfth intercostal, iliohypogastric, ilioinguinal nerves

OBLIQUUS INTERNUS ABDOMINIS

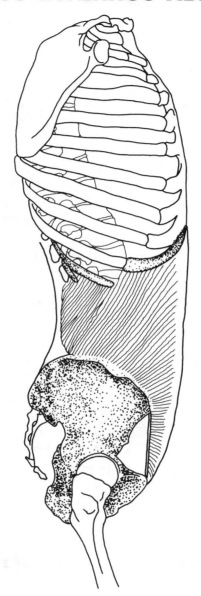

Trunk—lateral view

Origin	Lateral half of inguinal ligament, iliac crest, thoracolumbar fascia	**Action**	Compresses abdominal contents, laterally flexes and rotates vertebral column
Insertion	Cartilage of bottom three or four ribs, abdominal aponeurosis to linea alba	**Nerve**	Eighth to twelfth intercostal, iliohypogastric, ilioinguinal nerves

CREMASTER

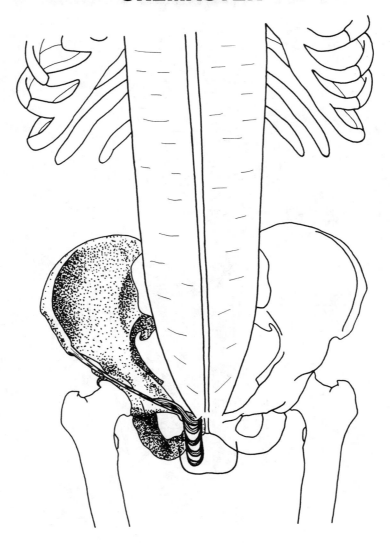

Trunk—anterior view

Origin	Inguinal ligament	**Action**	Pulls testes toward body
Insertion	Pubic tubercle, crest of pubis, sheath of rectus abdominis	**Nerve**	Genital branch of genitofemoral nerve

TRANSVERSUS ABDOMINIS

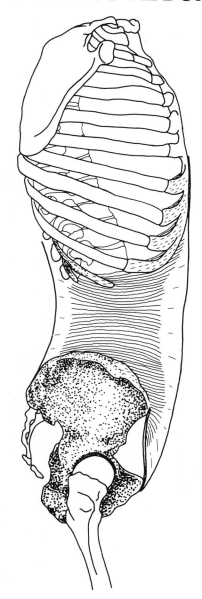

Trunk—lateral view

Origin   Lateral part of inguinal ligament, iliac crest, thoracolumbar fascia, cartilage of lower six ribs

Insertion Abdominal aponeurosis to linea alba

Action   Compresses abdomen

Nerve   Seventh to twelfth intercostal, iliohypogastric, ilioinguinal nerves

RECTUS ABDOMINIS

Origin	Crest of pubis, pubic symphysis
Insertion	Cartilage of fifth, sixth, and seventh ribs, xiphoid process
Action	Flexes vertebral column, compresses abdomen
Nerve	Seventh through twelfth intercostal nerves

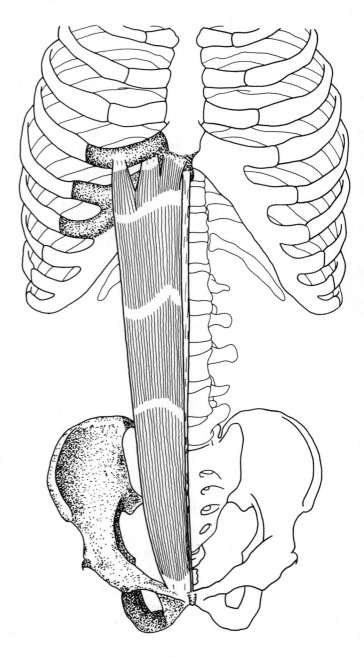

Trunk—anterior view

PYRAMIDALIS

Origin	Ventral surface of pubis, pubic ligament
Insertion	Linea alba
Action	Tenses linea alba
Nerve	T12

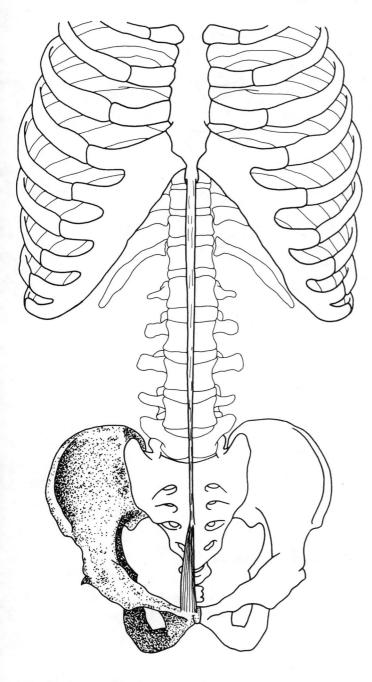

Trunk—anterior view

QUADRATUS LUMBORUM

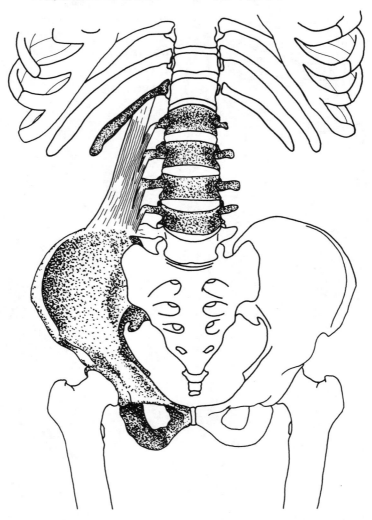

Lower trunk—anterior view

Origin	Iliolumbar ligament, iliac crest	**Action**	Laterally flexes vertebral column, fixes ribs for forced expiration
Insertion	Twelfth rib, transverse processes of upper four lumbar vertebrae	**Nerve**	T12, L1

CHAPTER SIX
MUSCLES OF THE SHOULDER AND ARM

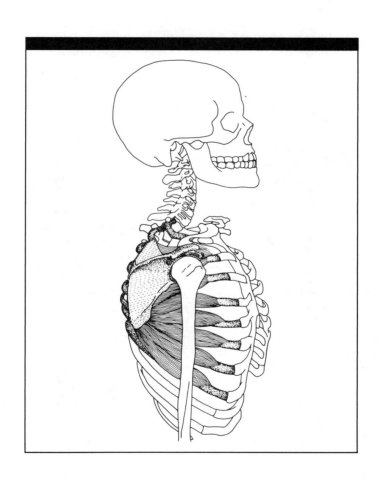

PECTORALIS MAJOR

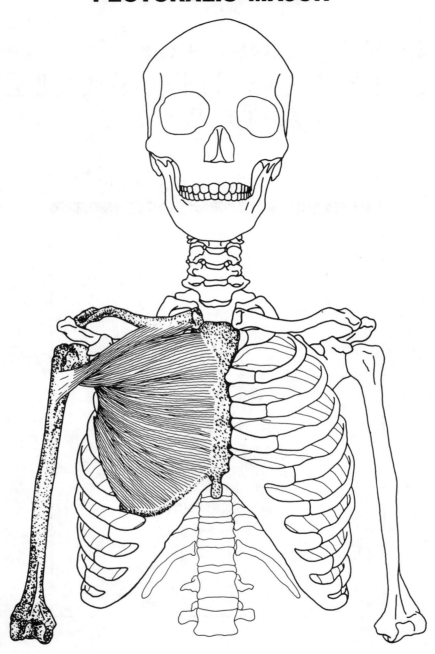

Anterior view

Origin	Medial half of the clavicle, sternum, first six costal cartilages, aponeurosis of the external oblique	**Action**	Adducts the arm, flexes arm, rotates the arm medially, depresses the arm and shoulder
Insertion	Lateral lip of intertubercular (bicipital) groove of humerus, crest below greater tubercle of the humerus	**Nerve**	Medial and lateral pectoral nerves (C5–C8, T1)

PECTORALIS MINOR

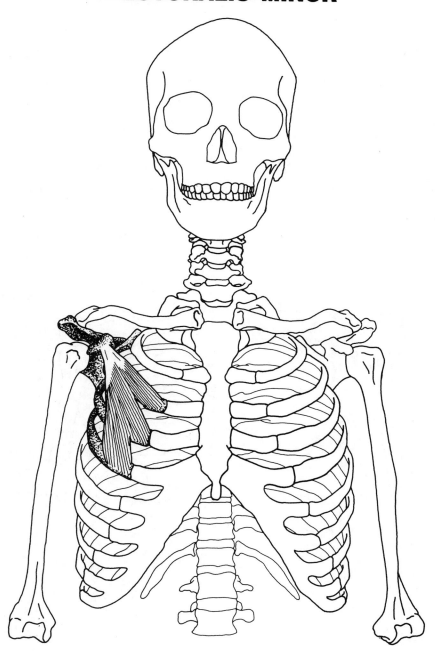

Anterior view

Origin	External surfaces of the third, fourth, and fifth ribs	**Action**	Draws scapula forward and downward, raises ribs in forced inspiration
Insertion	Coracoid process of the scapula		
		Nerve	Medial pectoral nerve (C8, T1)

SUBCLAVIUS

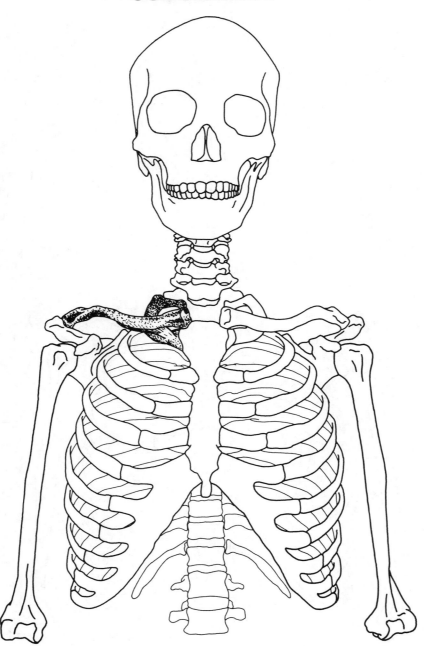

Anterior view

Origin	Junction of the first rib with its costal cartilage	**Action**	Depresses clavicle, draws shoulder forward and downward, steadies clavicle during movements of shoulder girdle
Insertion	Groove on the inferior (lower) surface of the clavicle	**Nerve**	C5, C6

CORACOBRACHIALIS

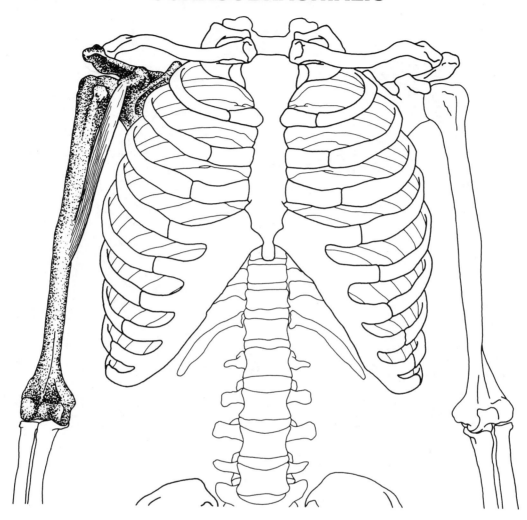

Anterior view

Origin	Tip (apex) of the coracoid process	**Action**	Flexes arm, weakly adducts arm
Insertion	Middle third of the medial surface and border of the humerus	**Nerve**	Musculocutaneous nerve (C6, C7)

BICEPS BRACHII

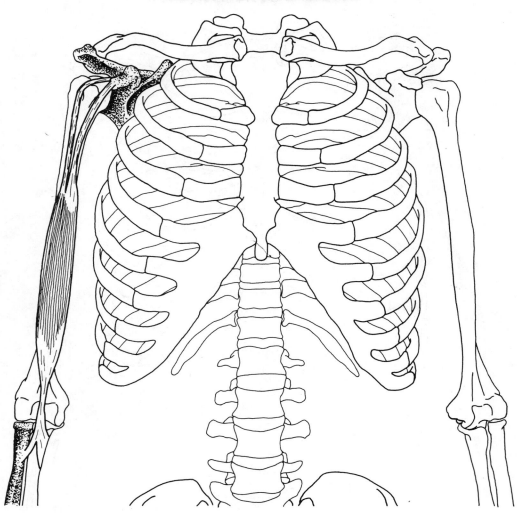

Anterior view

Origin	Long head—supraglenoid tubercle of scapula	**Action**	Supinates hand, flexes forearm, weak flexor of arm at shoulder joint
	Short head—coracoid process of scapula	**Nerve**	Musculocutaneous nerve (C5, C6)
Insertion	Tuberosity of radius, bicipital aponeurosis into deep fascia on medial part of forearm		

BRACHIALIS

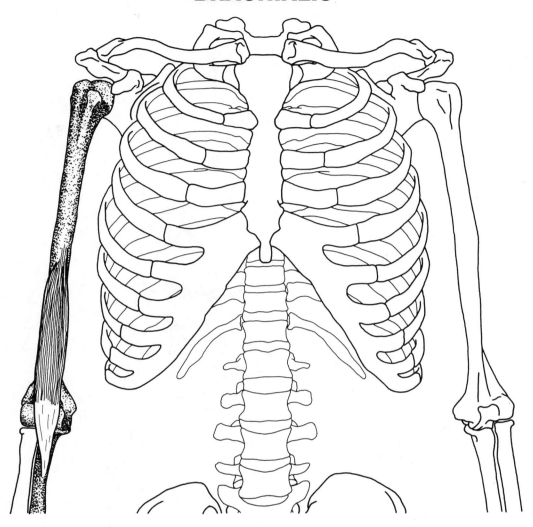

Anterior view

Origin	Anterior of lower half of humerus	**Action**	Flexes forearm
Insertion	Coronoid process of ulna, tuberosity of ulna	**Nerve**	Musculocutaneous nerve (C5, C6)

TRAPEZIUS

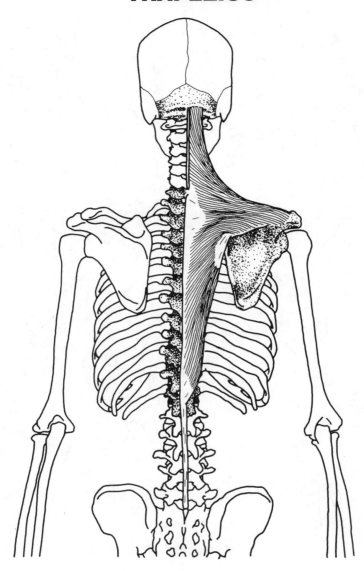

Posterior view

Origin	Medial third of superior nuchal line, external occipital protuberance, ligamentum nuchae, spinous processes and supraspinous ligaments of seventh cervical and all thoracic vertebrae	**Action**	Elevates lateral point of scapula (rotates scapula during abduction and elevation of arm), adducts scapula, lower portion depresses scapula
Insertion	Lateral third of clavicle, medial margin of acromion, entire length of spine of scapula	**Nerve**	Accessory (11th cranial), C3, C4

T12

LATISSIMUS DORSI

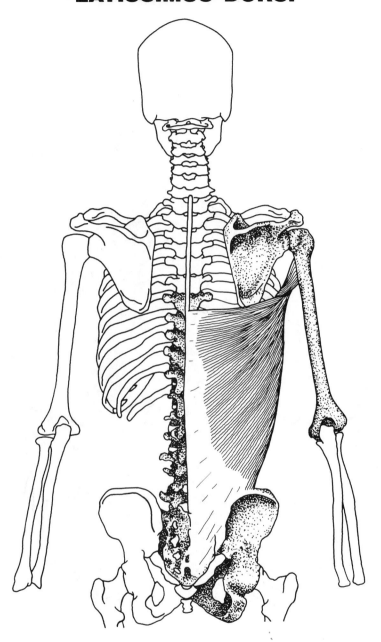

Posterior view

Origin	Spinous processes of the lower six thoracic vertebrae, lumbar vertebrae, sacral vertebrae, supraspinal ligament, and posterior part of the iliac crest through the lumbar (thoracolumbar) fascia, lower three or four ribs, inferior angle of the scapula	**Action**	Extends, adducts, and medially rotates the arm, draws the shoulder downward and backward, keeps inferior angle of scapula against the chest wall, accessory muscle of respiration
Insertion	Floor (bottom) of the bicipital groove of humerus	**Nerve**	Thoracodorsal nerve, (C6–C8)

LEVATOR SCAPULAE

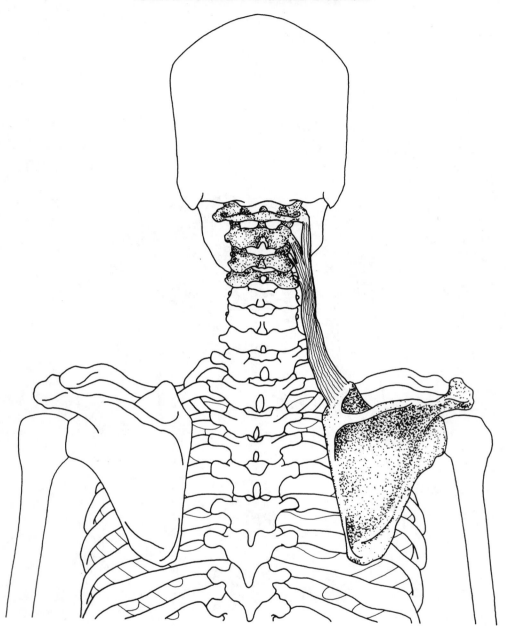

Posterior view

Origin	Posterior tubercles of the transverse processes of the first four cervical vertebrae	**Action**	Elevates medial border of scapula, rotates scapula to lower the lateral angle, acts with trapezius and rhomboids to pull scapula medially and upward, bends neck laterally
Insertion	Vertebral (medial) border of the scapula at and above the spine		
		Nerve	Dorsal scapular nerve (C5)

RHOMBOIDEUS MAJOR

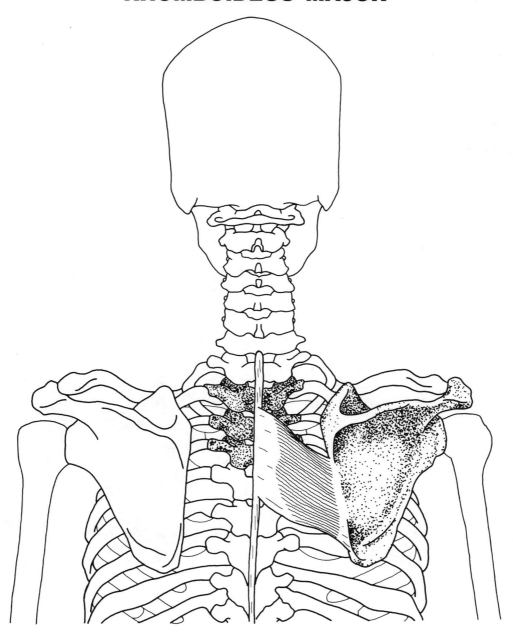

Posterior view

Origin   Spines of the second to fifth thoracic vertebrae, supraspinous ligament

Insertion   Medial border of the scapula below the spine

Action   Retracts and fixes scapula, elevates the medial border of the scapula, rotates the scapula to depress the lateral angle (assists in adduction of arm)

Nerve   Dorsal scapular nerve (C5)

RHOMBOIDEUS MINOR

*push hand
away from
pock*

*more
superior*

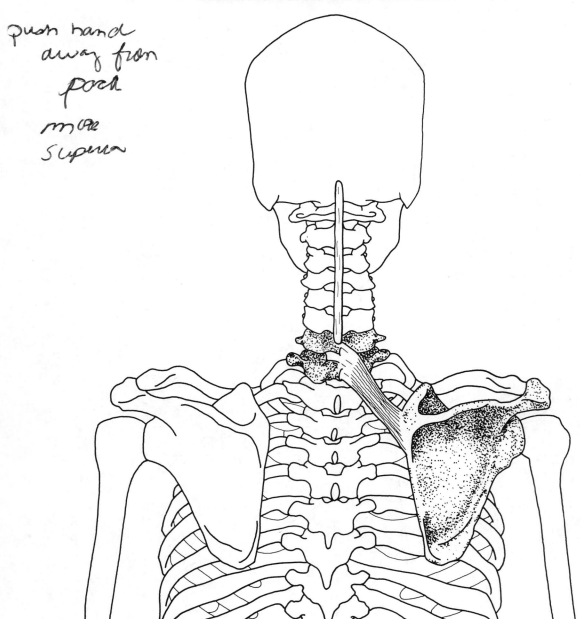

Posterior view

Origin   Spines of the seventh cervical and first thoracic vertebrae, lower part of the ligamentum nuchae

Insertion   Medial border of the scapula at the root of the spine

Action   Retracts and fixes scapula, elevates the medial border of the scapula, rotates the scapula to depress the lateral angle (assists in adduction of arm)

Nerve   Dorsal scapular nerve (C5)

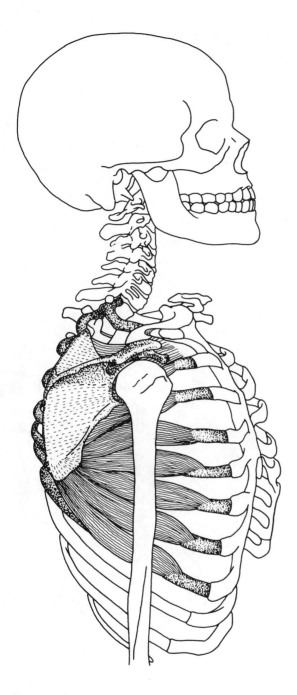

SERRATUS ANTERIOR

Origin   Outer surfaces and superior borders
of first eight or nine ribs, and fascia
covering first intercostal space

Insertion Anterior surface (costal surface) of the
medial border of the scapula

Action   Rotates and protracts scapula

Nerve   Long thoracic nerve (C5–C7)

Lateral view

DELTOIDEUS

Origin   Anterior portion—anterior border and
                    superior surface of the lateral third of
                    the clavicle

                    Middle portion—lateral border of the
                    acromion process

                    Posterior portion—lower border of the
                    crest of the spine of the scapula

Insertion   Deltoid tuberosity, on the middle of
                    the lateral surface of the shaft of the
                    humerus

Action   Abducts arm (middle portion), flexes
                    and medially rotates arm (anterior
                    portion), extends and laterally rotates
                    arm (posterior portion)

Nerve   Axillary nerve (C5, C6)

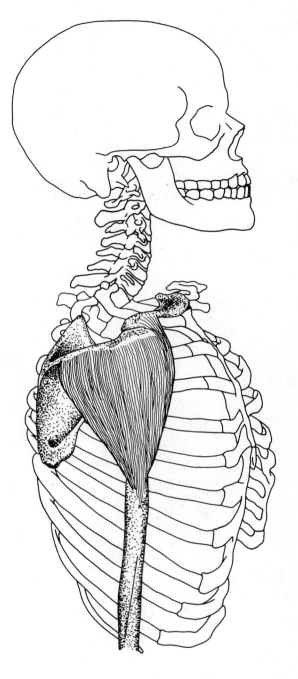

Lateral view

SUPRASPINATUS

Origin	Supraspinous fossa of scapula
Insertion	Upper part of the greater tuberosity of the humerus, capsule of the shoulder joint
Action	Aids the deltoid in abduction of the arm, draws humerus toward the glenoid fossa and strengthens the shoulder joint (rotator cuff), weak lateral rotator and flexor
Nerve	Suprascapular nerve (C5)

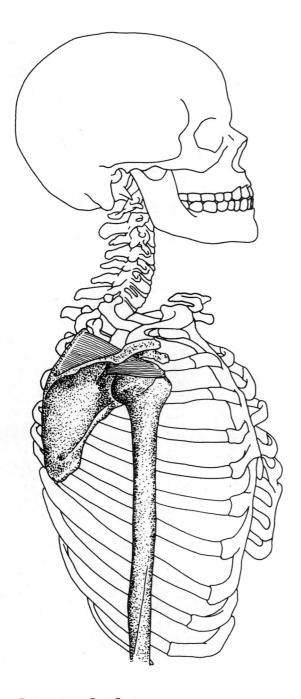

Lateral view

INFRASPINATUS

Origin Infraspinous fossa of the scapula

Insertion Middle facet of the greater tuberosity of the humerus, capsule of the shoulder joint

Action Laterally rotates the arm, upper part abducts arm, lower part adducts arm, draws the humerus toward the glenoid fossa (rotator cuff)

Nerve Suprascapular nerve (C5, C6)

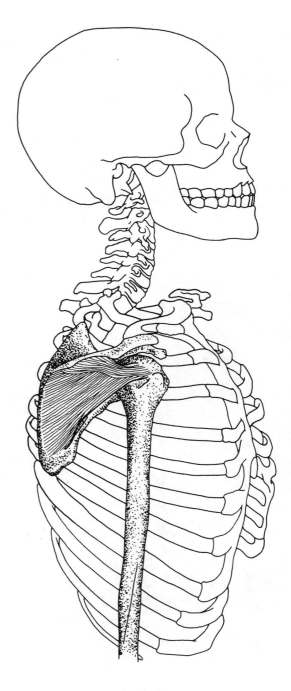

Lateral view

TERES MINOR

Origin	Upper two-thirds of the dorsal surface of the axillary border of the scapula
Insertion	The capsule of the shoulder joint, the lower facet of the greater tuberosity of the humerus
Action	Laterally rotates arm, weakly adducts arm, draws humerus toward glenoid fossa, stabilizes the shoulder joint (rotator cuff)
Nerve	Axillary nerve (C5)

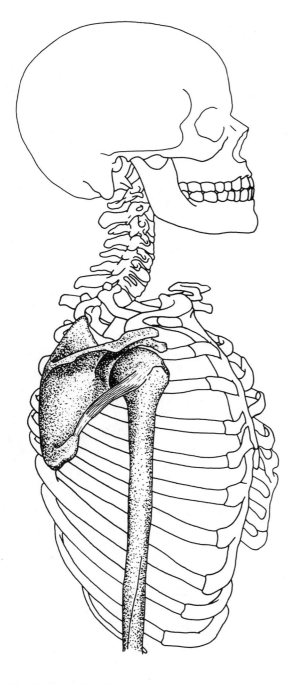

Lateral view

SUBSCAPULARIS

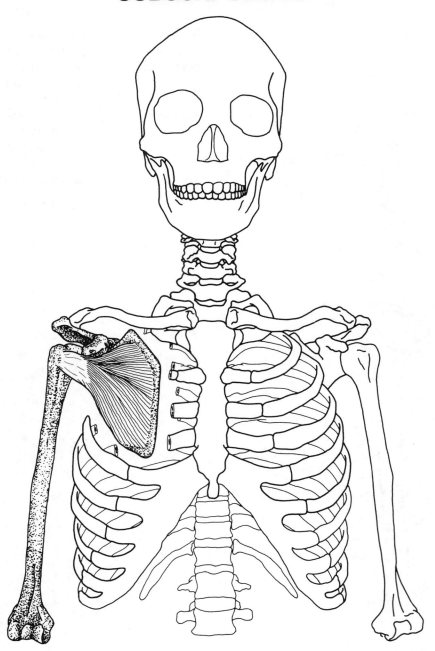

Anterior view

(Upper ribs cut away)

Origin	Subscapular fossa on the anterior surface of scapula	**Action**	Medially rotates arm, stabilizes shoulder joint (rotator cuff), assists in both flexion and extension and abduction and adduction, depending upon position of the arm
Insertion	Lesser tuberosity of the humerus, ventral part of the capsule of the shoulder joint		
		Nerve	Upper and lower subscapular nerves (C5, C6)

TERES MAJOR

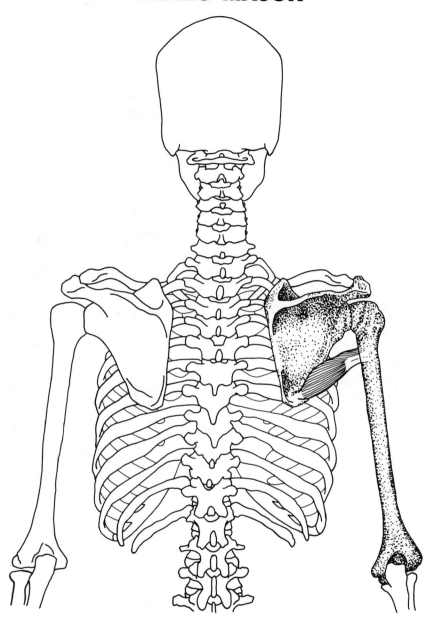

Posterior view

Origin   Lower third of the posterior surface of
             the lateral border of the scapula, near
             the inferior angle

Insertion   Medial lip of the bicipital groove of the
             humerus

Action   Medially rotates arm, adducts arm,
             extends arm

Nerve   Lower subscapular nerve (C5, C6)

TRICEPS BRACHII

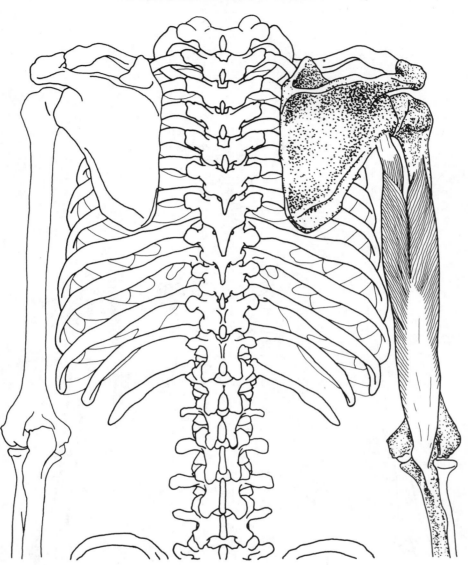

Posterior view

Origin

Long head—infraglenoid tubercle of the scapula

Lateral head—upper half of the posterior surface of the shaft of the humerus

Medial head—posterior surface of the lower half of the shaft of the humerus

Insertion

Posterior part of olecranon process of the ulna

Action

Extends forearm, long head aids in adduction if arm is abducted

Nerve

Radial nerve (C7, C8)

ANCONEUS

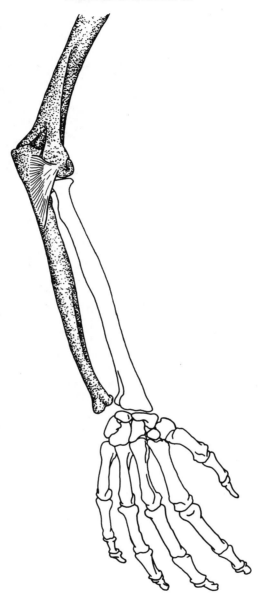

Posterior view of arm

Origin   Posterior part of lateral epicondyle of
            the humerus

Insertion   Lateral surface of the olecranon
            process and posterior surface of ulna

Action   Extends arm (assists triceps)

Nerve   Radial nerve (C7, C8)

CHAPTER SEVEN
MUSCLES OF THE FOREARM AND HAND

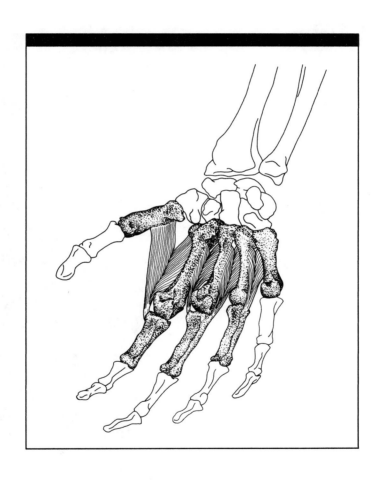

PRONATOR TERES

Origin   Humeral head—medial supracondylar ridge and medial epicondyle of the humerus

                        Ulnar head—medial border of the coronoid process of the ulna

Insertion   Middle of lateral surface of the radius (pronator tuberosity)

Action   Pronates and flexes the forearm and hand

Nerve   Median nerve (C6, C7)

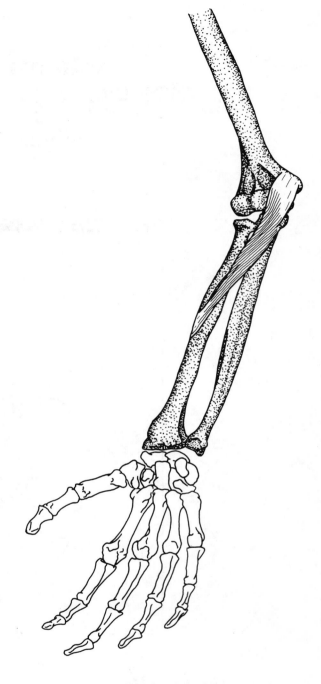

Forearm—anterior view

FLEXOR CARPI RADIALIS

Origin	Medial epicondyle of the humerus through the common tendon
Insertion	Front of the bases of the second and third metacarpal bones
Action	Flexes the hand, abducts the hand at the wrist joint
Nerve	Median nerve (C6, C7)

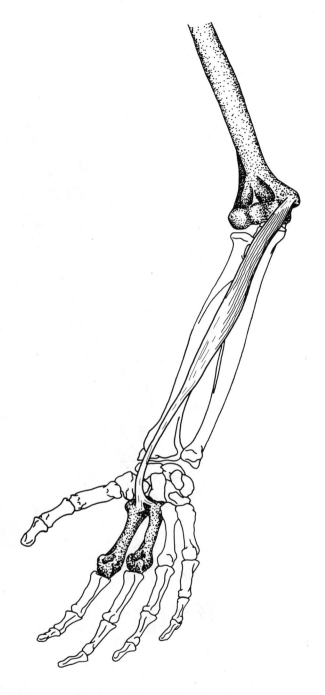

Forearm—anterior view

PALMARIS LONGUS

Origin	Medial epicondyle of the humerus through the common tendon
Insertion	Front (central part) of the flexor retinaculum and apex of the palmar aponeurosis
Action	Flexes the hand
Nerve	Median nerve (C6, C7)

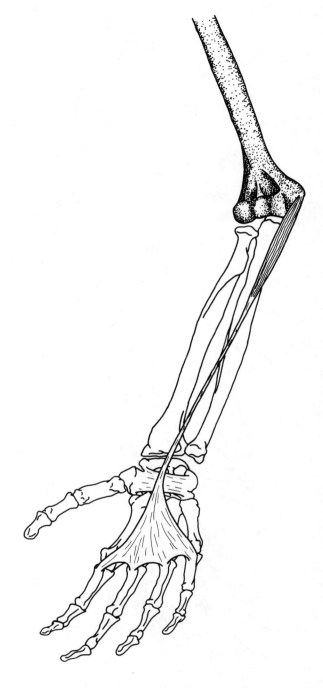

Forearm—anterior view

FLEXOR CARPI ULNARIS

Origin	Humeral head—medial epicondyle of the humerus through the common tendon
	Ulnar head—medial margin of olecranon process of ulna, dorsal border of shaft of the ulna
Insertion	Pisiform bone, hook of the hamate and base of the fifth metacarpal bone
Action	Flexes and adducts the hand at the wrist joint
Nerve	Ulnar nerve (C8, T1)

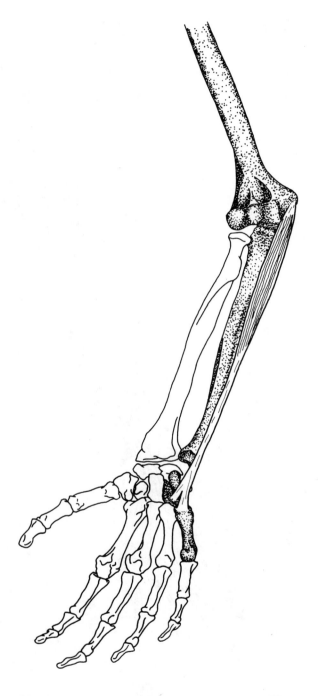

Forearm—anterior view

FLEXOR DIGITORUM SUPERFICIALIS

Origin   Humeroulnar head—medial
epicondyle of the humerus through
common tendon, medial margin of the
coronoid process of ulna
Radial head—anterior surface of shaft
of radius

Insertion   Four tendons divide into two slips
each, slips insert into the sides
(margins of the anterior surfaces) of
the middle phalanges of four fingers

Action   Flexes the middle phalanges of the
fingers

Nerve   Median nerve (C7, C8, T1)

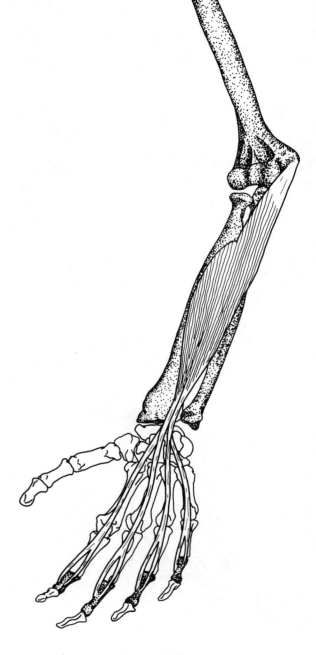

Forearm—anterior view

FLEXOR DIGITORUM PROFUNDUS

Origin Upper three-fourths of anterior and medial surfaces of shaft of ulna and medial side of the coronoid process, interosseous membrane

Insertion Front of base of distal phalanges of medial four fingers

Action Flexes distal phalanges

Nerve Ulnar nerve supplies the medial half of the muscle (going to the little and ring fingers) (C8, T1)

Anterior interosseous branch of median nerve supplies lateral half (going to index and middle fingers)

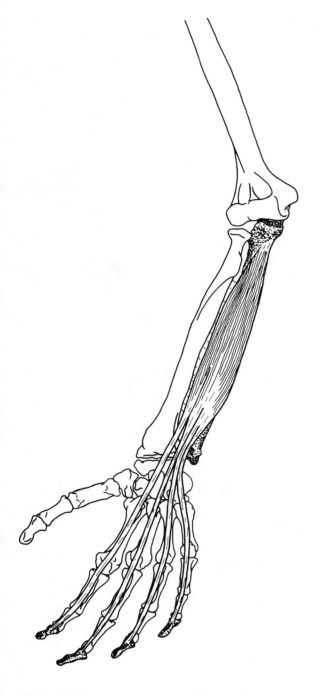

Forearm—anterior view

FLEXOR POLLICIS LONGUS

Origin	Middle of anterior surface of shaft of radius, interosseous membrane, medial epicondyle of humerus and often coronoid process of ulna
Insertion	Palmar aspect of base of the distal phalanx of thumb
Action	Flexes distal phalanx of the thumb
Nerve	Anterior interosseous branch of median nerve (C8, T1)

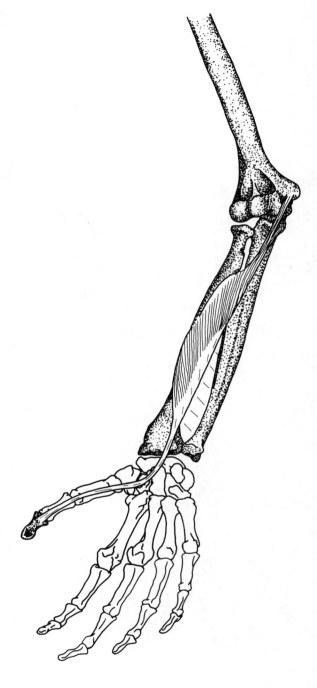

Forearm—anterior view

PRONATOR QUADRATUS

Origin	Anterior surface of distal part of shaft of ulna
Insertion	Lower portion of anterior surface of shaft of radius, distal part of lateral border of radius
Action	Pronates forearm and hand
Nerve	Anterior interosseous branch of median nerve (C8, T1)

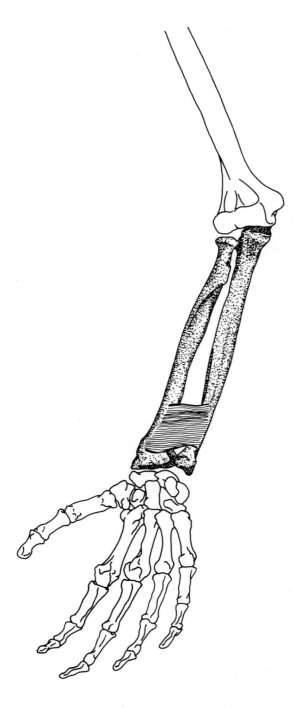

Forearm—anterior view

BRACHIORADIALIS

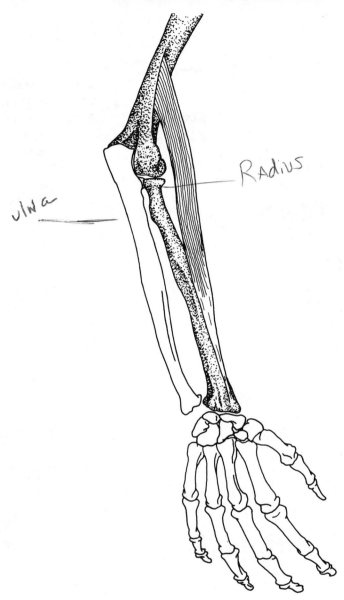

Radius

ulna

Forearm—dorsal view

Origin	Upper two-thirds of lateral supracondylar ridge of humerus	**Action**	Flexes forearm at elbow joint
Insertion	Base of styloid process and lateral surface of radius	**Nerve**	Radial nerve (C5, C6)

EXTENSOR CARPI RADIALIS LONGUS

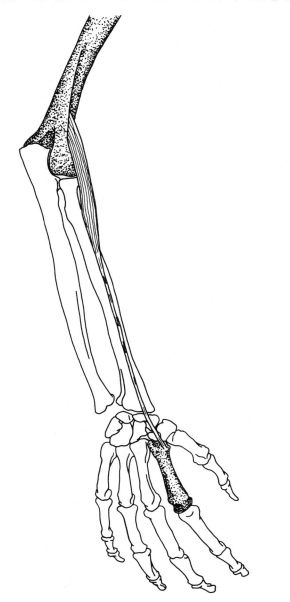

403
7179

Forearm—dorsal view

Origin	Lower third of lateral supracondylar ridge of humerus	**Action**	Extends and abducts hand at wrist joint
Insertion	Posterior surface of the base of the second metacarpal bone	**Nerve**	Radial nerve (C6, C7)

EXTENSOR CARPI RADIALIS BREVIS

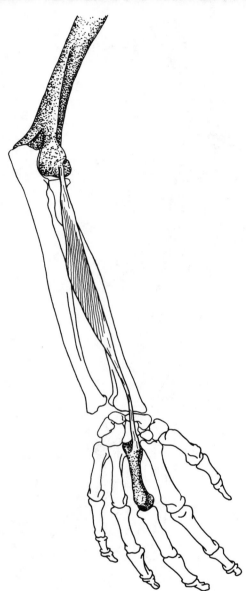

Forearm—dorsal view

Origin	Lateral epicondyle of humerus	**Action**	Extends hand, assists in abducting hand
Insertion	Dorsal surface of third metacarpal bone	**Nerve**	Radial nerve (C6, C7)

EXTENSOR DIGITORUM COMMUNIS

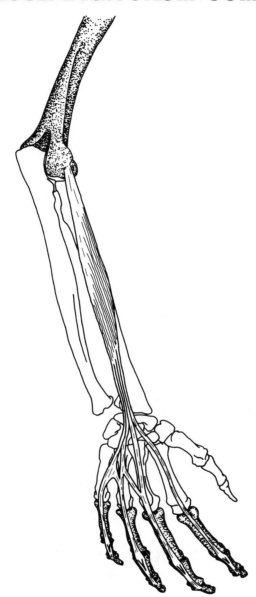

Forearm and hand-dorsal view

Origin   Common tendon attached to lateral epicondyle of humerus

Insertion   Lateral and dorsal surfaces of all the phalanges of the four fingers

Action   Extends the fingers and wrist

Nerve   Deep branch of radial nerve (C6–C8)

EXTENSOR DIGITI MINIMI

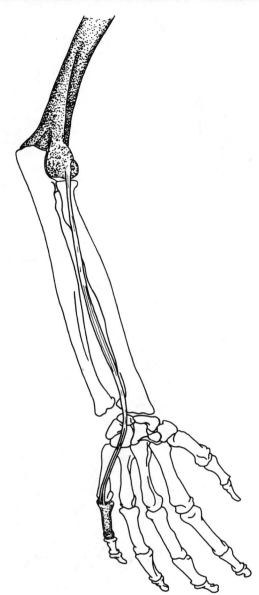

Forearm and hand—dorsal view

Origin	Common tendon attached to lateral epicondyle of humerus	**Action**	Extends little finger
		Nerve	Radial nerve (C6–C8)
Insertion	Dorsum of first phalanx of little finger		

EXTENSOR CARPI ULNARIS

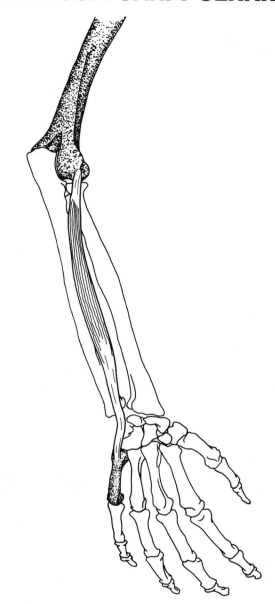

Forearm and hand-dorsal view

Origin   Common tendon attached to lateral epicondyle of humerus

Insertion   Posterior surface of base of fifth metacarpal bone

Action   Extends and adducts hand at wrist joint

Nerve   Radial nerve (C6–C8)

SUPINATOR

Origin	Lateral epicondyle of humerus, lateral ligament (radial collateral) of elbow, annular ligament of superior radioulnar joint supinator crest of ulna
Insertion	Dorsal and lateral surfaces of upper third of radius
Action	Supinates the forearm and hand
Nerve	Radial nerve (C6)

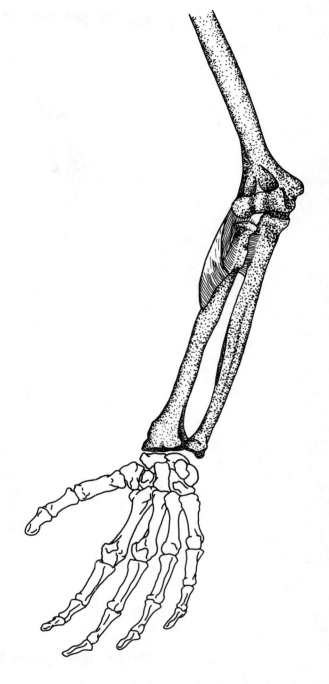

Forearm and hand—anterior view

ABDUCTOR POLLICIS LONGUS

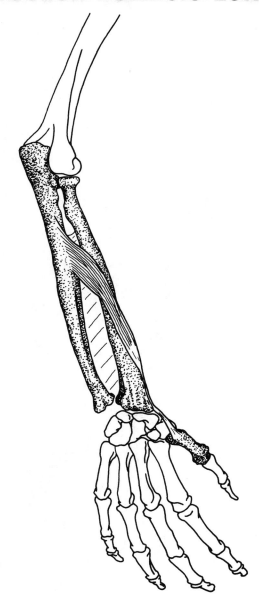

Forearm and hand—dorsal view

Origin Posterior (dorsal) surface of shaft of radius, ulna, interosseous membrane

Insertion Posterior surface of base of first metacarpal bone

Action Abducts thumb, abducts wrist

Nerve Radial nerve (C6, C7)

EXTENSOR POLLICIS BREVIS

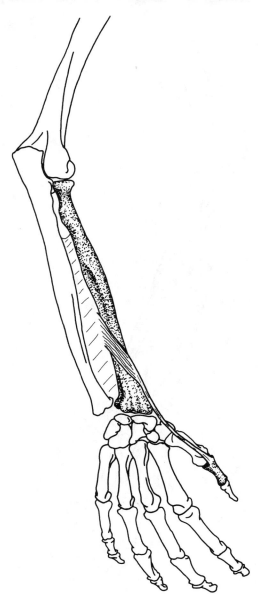

Forearm and hand—dorsal view

Origin	Posterior surface of radius, adjacent part of interosseous membrane	**Action**	Extends proximal phalanx of thumb, abducts hand
Insertion	Base of proximal phalanx of thumb	**Nerve**	Radial nerve (C6, C7)

EXTENSOR POLLICIS LONGUS

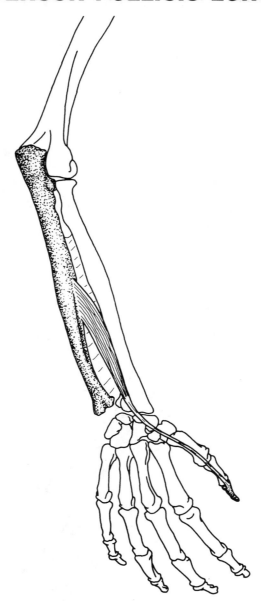

Forearm and hand—dorsal view

Origin Middle third of dorsal surface of ulna, **Action** Extends thumb, abducts hand
     interosseous membrane   **Nerve** Radial nerve (C6–C8)

Insertion Base of distal phalanx of thumb

EXTENSOR INDICIS

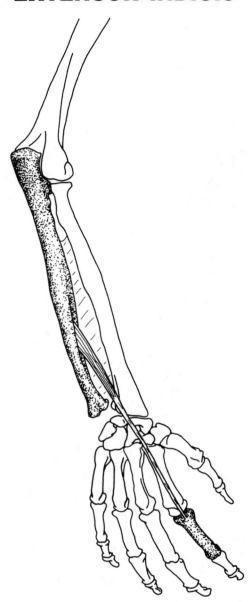

Forearm and hand—dorsal view

Origin	Posterior surface of ulna and adjacent part of interosseous membrane	**Action**	Extends index finger
Insertion	Extensor expansion on back of index finger	**Nerve**	Radial nerve (C6–C8)

PALMARIS BREVIS

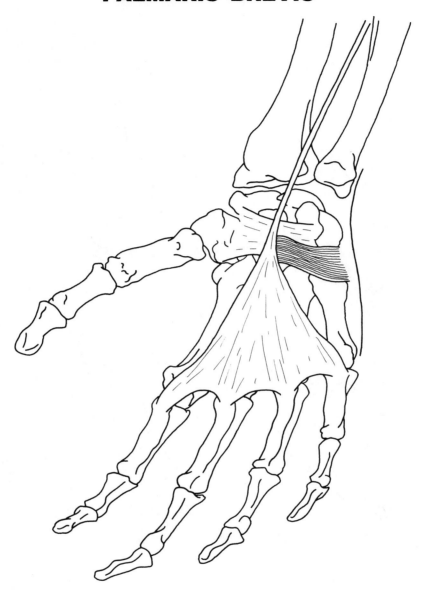

Hand—palmar view

Origin	Flexor retinaculum, palmar aponeurosis	**Action**	Corrugates skin of palm
		Nerve	Ulnar nerve (C8)
Insertion	Skin of the palm		

ABDUCTOR POLLICIS BREVIS

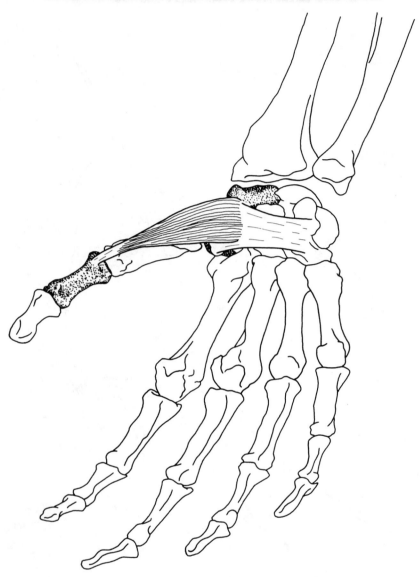

Hand—palmar view

Origin   Tubercle of scaphoid, tubercle of   **Action**   Abducts thumb and moves it
             trapezium, flexor retinaculum,   anteriorly
             transverse carpal ligament   **Nerve**   Median (C6, C7)

Insertion   Base of proximal phalanx of thumb

FLEXOR POLLICIS BREVIS

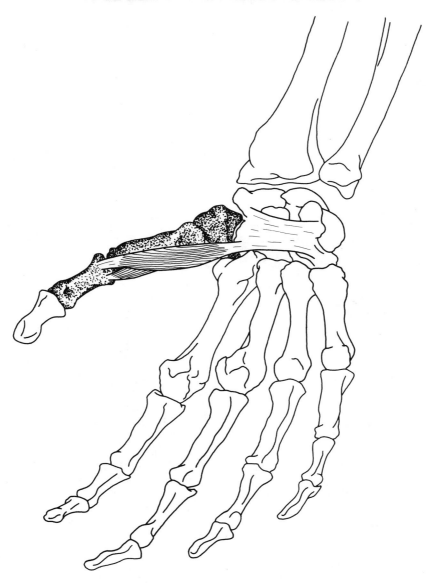

Hand—palmar view

Origin	Flexor retinaculum and trapezium, and first metacarpal bone	**Nerve**	Lateral portion—median nerve (C6, C7) Medial portion—ulnar nerve (C8, T1)
Insertion	Base of proximal phalanx of thumb		
Action	Flexes metacarpophalangeal joint of thumb, assists in abduction and rotation of thumb		

OPPONENS POLLICIS

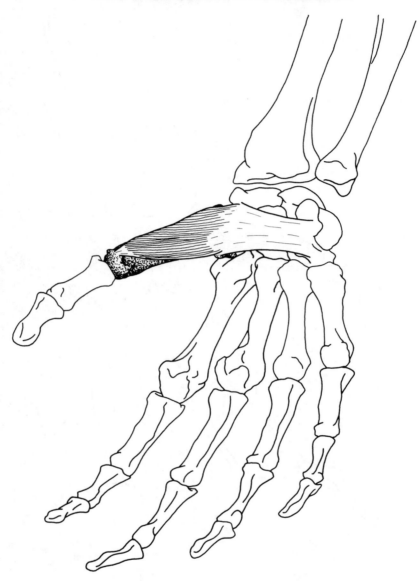

Hand—palmar view

Origin	Flexor retinaculum, tubercle of trapezium	**Action**	Rotates thumb into opposition with fingers
Insertion	Lateral border of first metacarpal bone	**Nerve**	Median nerve (C6, C7)

ADDUCTOR POLLICIS

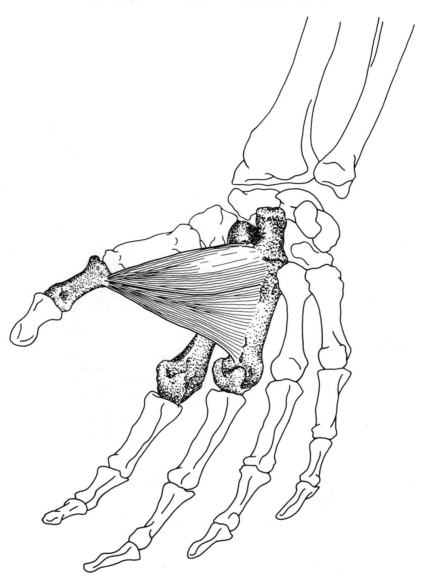

Hand—palmar view

Origin

Oblique head—anterior surfaces of second and third metacarpals, capitate, trapezoid
Transverse head—anterior surface of third metacarpal bone

Insertion

Medial side of base of proximal phalanx of the thumb

Action

Adducts thumb

Nerve

Ulnar nerve (C8, T1)

ABDUCTOR DIGITI MINIMI

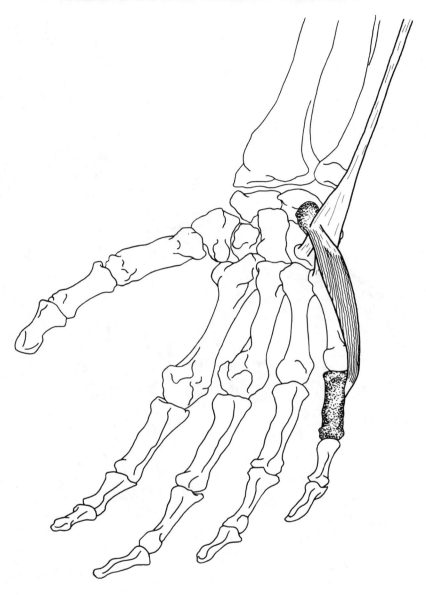

Hand—palmar view

Origin   Pisiform bone, tendon of flexor carpi
                  ulnaris

Insertion   Medial side of base of proximal
                  phalanx of little finger

Action   Abducts little finger

Nerve   Ulnar nerve (C8, T1)

FLEXOR DIGITI MINIMI

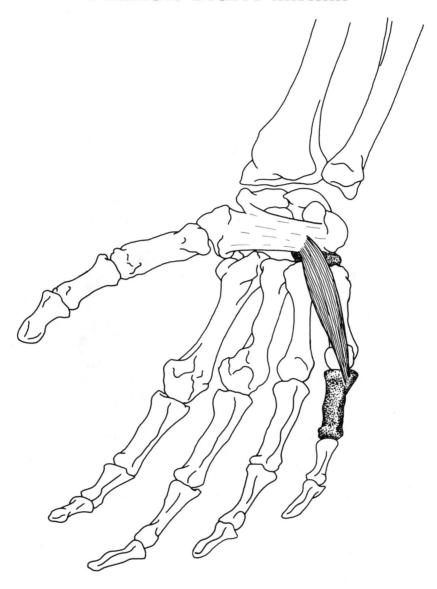

Hand—palmar view

Origin	Anterior surface of flexor retinaculum, hook of hamate	**Action**	Flexes little finger at metacarpophalangeal joint
Insertion	Medial side of base of proximal phalanx of little finger	**Nerve**	Ulnar nerve (C8, T1)

OPPONENS DIGITI MINIMI

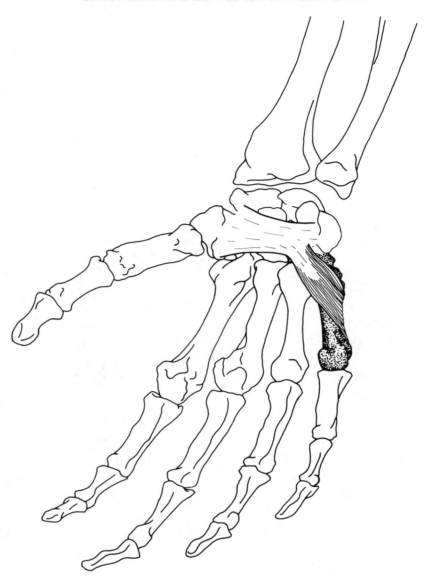

Hand—palmar view

Origin	Anterior surface of flexor retinaculum, hook of hamate	**Action**	Rotates fifth metacarpal bone, draws fifth metacarpal bone forward, assists flexor digiti minimi in flexing carpometacarpal joint of little finger
Insertion	Whole length of medial border of fifth metacarpal bone		
		Nerve	Ulnar nerve (C8, T1)

LUMBRICALES*
(Four muscles)

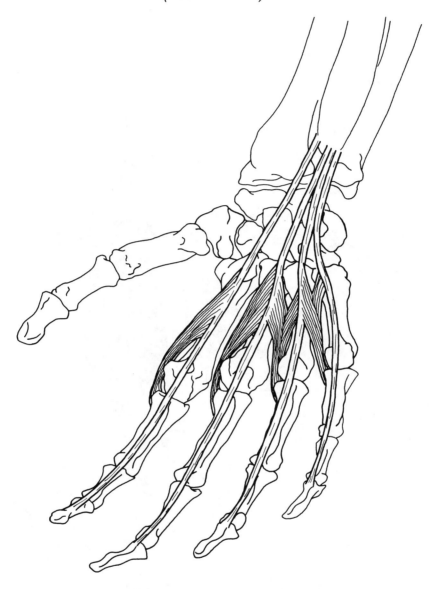

Hand—palmar view

Origin	Tendons of flexor digitorum profundus in palm	**Nerve**	Lateral lumbricals (first and second)— median nerve (C6, C7)
Insertion	Lateral side of corresponding tendon of extensor digitorum on fingers		Medial lumbricals (third and fourth)— ulnar nerve (C8)
Action	Flex metacarpophalangeal joints, extend interphalangeal joints		*Associated with the tendons of flexor digitorum profundus

PALMAR INTEROSSEI

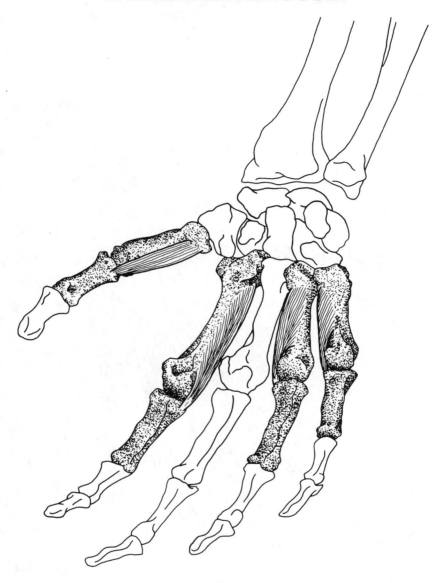

Hand—palmar view

Origin

First—medial side of base of first metacarpal bone

Second, third, and fourth—anterior surfaces of second, fourth, and fifth metacarpal bones

Insertion

First—medial side of base of proximal phalanx of thumb

Second—medial side of base of proximal phalanx of index finger

Third and fourth—lateral side of proximal phalanges of ring finger and little finger

Action

Adducts fingers toward center of third finger at metacarpophalangeal joints

Nerve

Ulnar nerve (C8, T1)

DORSAL INTEROSSEI

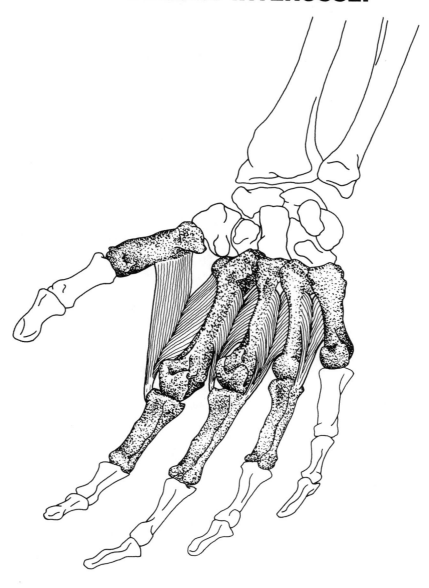

Hand—palmar view

Origin	By two heads from adjacent sides of first and second, second and third, third and fourth, and fourth and fifth metacarpal bones		Third—medial side of base of proximal phalanx of middle finger Fourth—medial side of base of proximal phalanx of ring finger
Insertion	First—lateral side of base of proximal phalanx of index finger Second—lateral side of base of proximal phalanx of middle finger	**Action**	Abduct fingers away from center of third finger at metacarpophalangeal joints
		Nerve	Ulnar nerve (C8, T1)

CHAPTER EIGHT
MUSCLES OF THE HIP AND THIGH

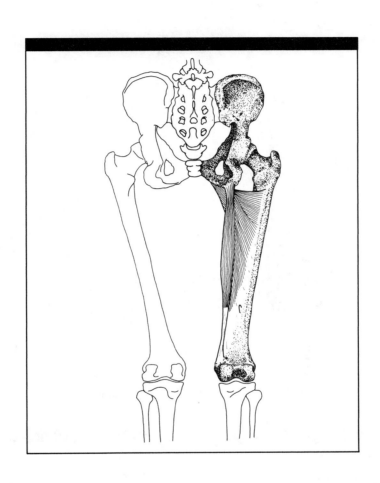

PSOAS MAJOR

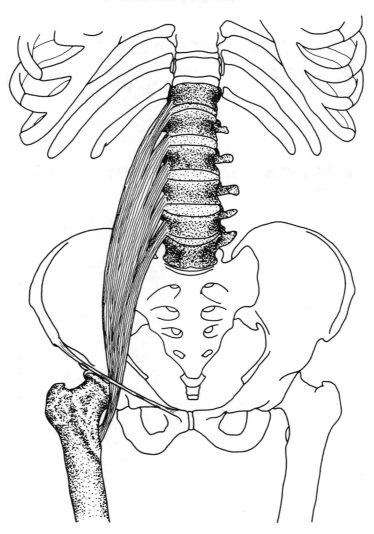

Lumbar region, hip, and thigh—anterior view

Origin	Bases of transverse processes of all lumbar vertebrae, bodies of twelfth thoracic and all lumbar vertebrae, intervertebral disks above each lumbar vertebra	**Insertion**	Lesser trochanter of femur
		Action	Flexes thigh at hip joint, flexes vertebral column
		Nerve	Branches from lumbar plexus (L2, L3) and sometimes L1 or L4

PSOAS MINOR
(Sometimes absent)

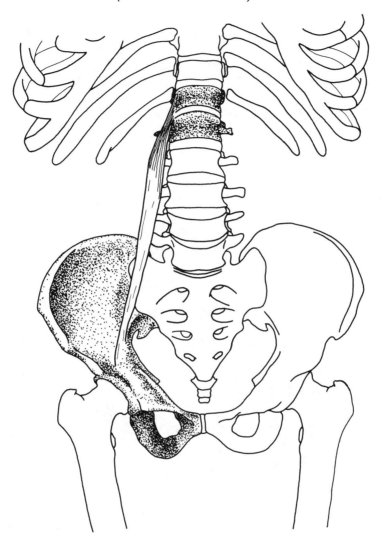

Lumbar region, hip, and thigh—anterior view

Origin	Sides of bodies of twelfth thoracic and first lumbar vertebrae	**Action**	Flexes pelvis and lumbar vertebral column
Insertion	Arcuate line to iliopectineal eminence	**Nerve**	First lumbar nerve from lumbar plexus

ILIACUS

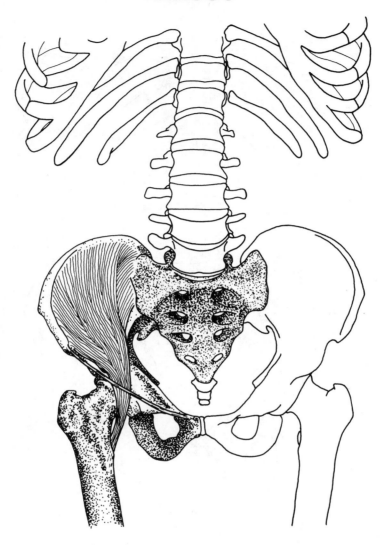

Lumbar region, hip, and thigh-anterior view

Origin   Upper two-thirds of iliac fossa, ala of sacrum and adjacent ligaments, anterior inferior iliac spine

Insertion   Onto tendon of psoas major, which continues into lesser trochanter of femur (together the two muscles form the iliopsoas)

Action   Flexes thigh at hip joint

Nerve   Femoral nerve (L2, L3)

PIRIFORMIS

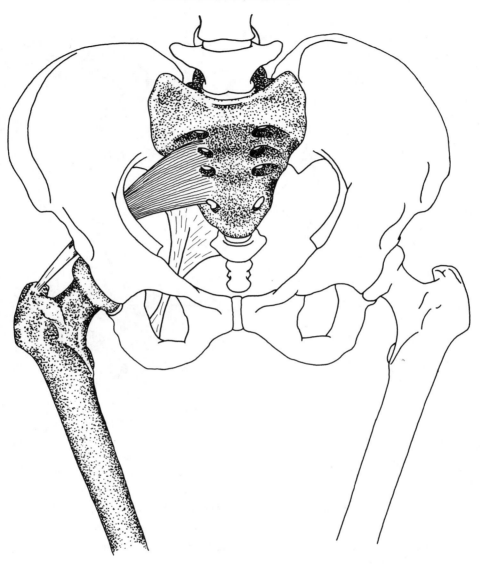

Hip and thigh—anterior view

Origin	Internal surface of sacrum, sacrotuberous ligament	**Action**	Laterally rotates thigh at hip joint, abducts thigh
Insertion	Upper border of greater trochanter	**Nerve**	Anterior rami of first and second sacral nerves

OBTURATOR INTERNUS

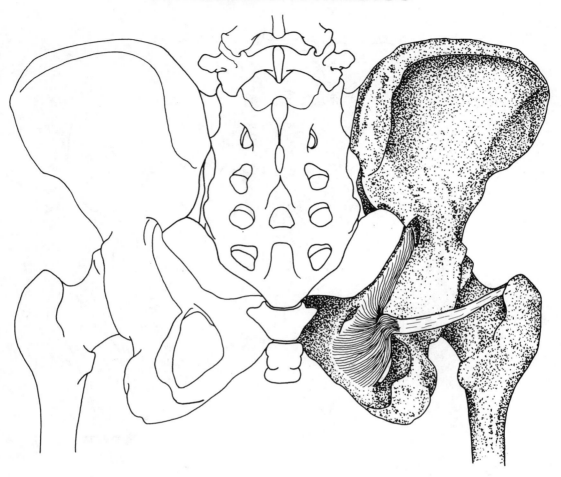

Hip—posterior view

Origin   Pelvic surface of obturator membrane and surrounding bones (ilium, ischium, pubis)

Insertion   Common tendon with superior and inferior gemelli to medial surface of greater trochanter

Action   Laterally rotates thigh at hip joint

Nerve   Nerve from sacral plexus (L5, S1–S3)

GEMELLUS SUPERIOR

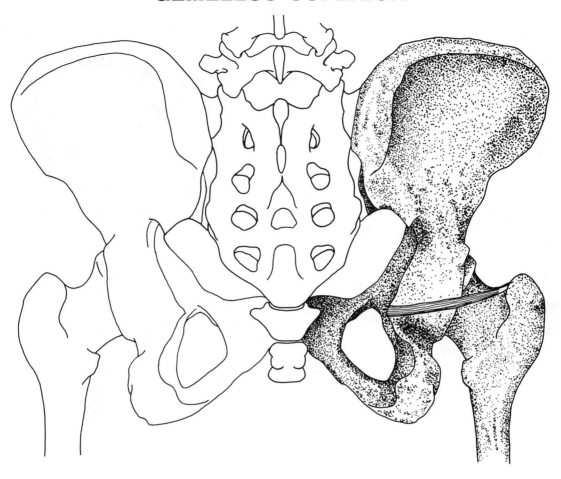

Hip—posterior view

Origin	Spine of ischium	**Action**	Laterally rotates thigh at hip joint
Insertion	With tendon of obturator internus into upper border of greater trochanter	**Nerve**	Branch of nerve to obturator internus from sacral plexus

GEMELLUS INFERIOR

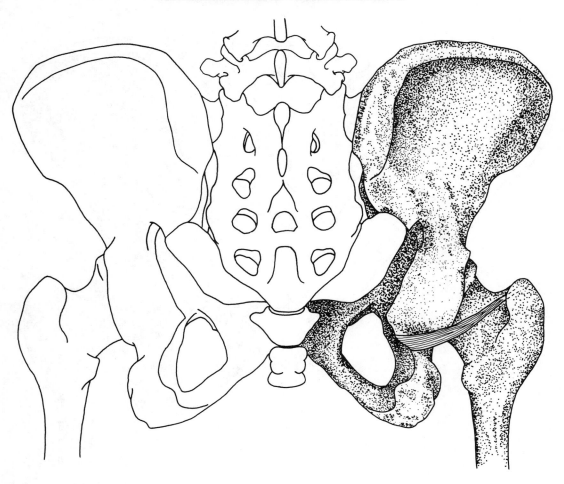

Hip—posterior view

Origin   Upper margin of ischial tuberosity

Insertion   With tendon of obturator internus into
                upper border of greater trochanter

Action   Laterally rotates thigh at hip joint

Nerve   Branch of nerve to quadratus femoris
                from sacral plexus

OBTURATOR EXTERNUS

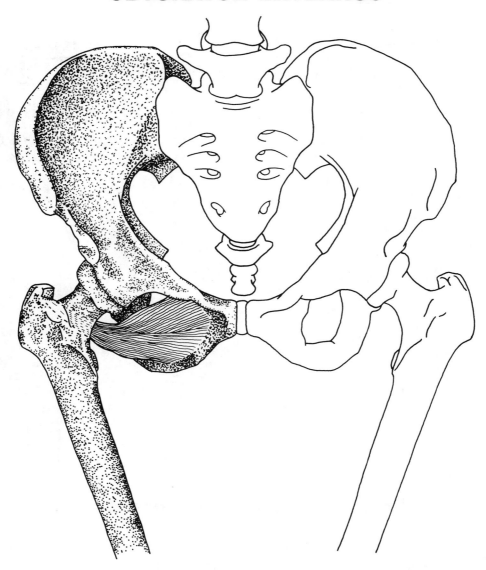

Hip and thigh—anterior view

Origin
Outer surface of superior and inferior rami of pubis and ramus of ischium surrounding obturator foramen

Insertion
Trochanteric fossa of femur

Action
Laterally rotates thigh

Nerve
Obturator nerve (L3, L4)

QUADRATUS FEMORIS

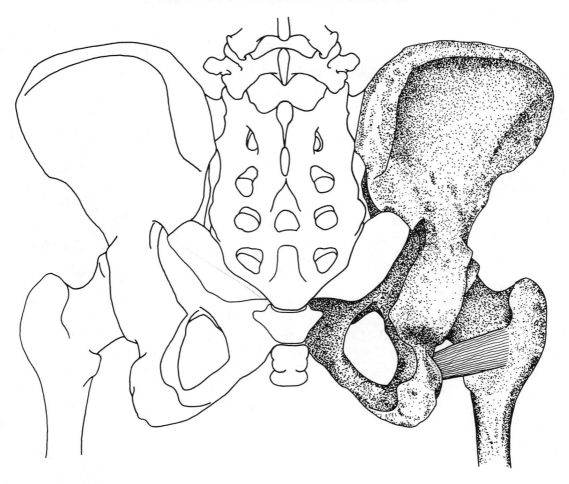

Hip and thigh—posterior view

Origin	Lateral border of ischial tuberosity	**Action**	Laterally rotates thigh at hip joint
Insertion	Below intertrochanteric crest (quadrate line)	**Nerve**	Branch from sacral plexus (L5, S1)

GLUTEUS MAXIMUS

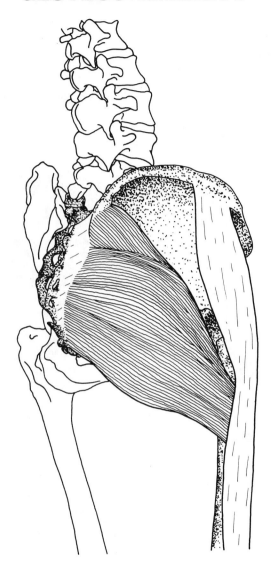

Hip and thigh—lateral view

Origin

Outer surface of ilium behind posterior gluteal line, adjacent posterior surface of sacrum and coccyx, sacrotuberous ligament, aponeurosis of erector spinae (sacrospinalis)

Insertion

Iliotibial tract of fascia lata, gluteal tuberosity of femur

Action

Extends and laterally rotates hip joint, extends trunk

Nerve

Inferior gluteal nerve (L5, S1, S2)

Not much of Lever arm ~ Bring toe out for me

GLUTEUS MEDIUS

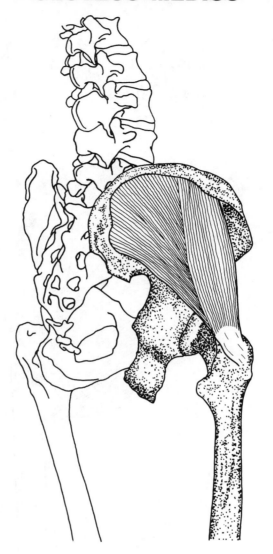

Hip and thigh—lateral view

Origin	Outer surface of ilium inferior to iliac crest	**Action**	Abducts femur at hip joint and rotates thigh medially
Insertion	Lateral surface of greater trochanter	**Nerve**	Superior gluteal nerve (L4, L5, S1)

GLUTEUS MINIMUS

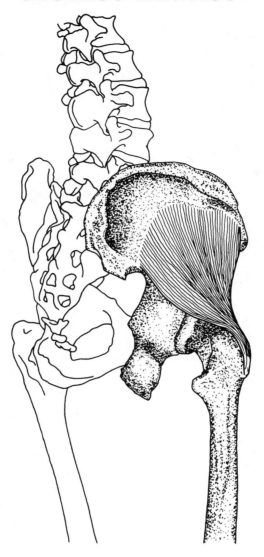

Hip and thigh—lateral view

Origin	Outer surface of ilium between middle (anterior) and inferior gluteal lines	**Action**	Abducts femur at hip joint and rotates thigh medially
Insertion	Anterior surface of greater trochanter	**Nerve**	Superior gluteal nerve (L4, L5, S1)

TENSOR FASCIAE LATAE

Origin   Outer edge of iliac crest between
                    anterior superior iliac spine and iliac
                    tubercle

Insertion   Iliotibial tract on upper part of thigh

Action   Flexes and medially rotates thigh

Nerve   Superior gluteal nerve (L4, L5, S1)

Amulation - walking

Assist adduct

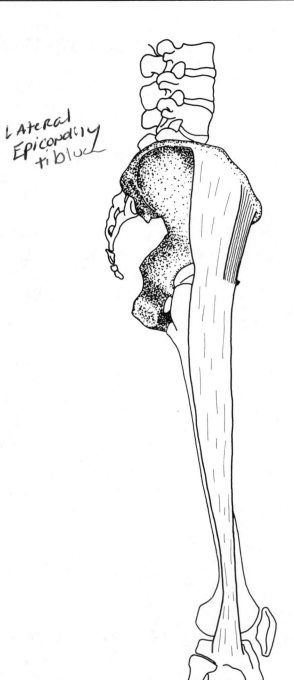

*Lateral
Epicondily
tibluc*

Hip and thigh—lateral view

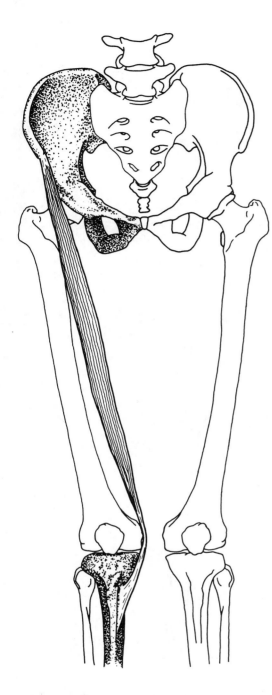

SARTORIUS

Origin	Anterior superior iliac spine and area immediately below it
Insertion	Upper part of medial surface of shaft of tibia
Action	Flexes, abducts, and laterally rotates thigh at hip joint, flexes and slightly medially rotates leg at knee joint after flexion
Nerve	Femoral nerve (L2, L3)

ASIS
Shaft of
tibia

Hip, thigh, and leg—anterior view

RECTUS FEMORIS
(One of quadriceps femoris)

Origin	Anterior head—anterior inferior iliac spine Posterior head—ilium above acetabulum
Insertion	Patella then by patellar ligament to tuberosity of the tibia
Action	Extends leg at knee joint, flexes thigh at hip joint
Nerve	Femoral nerve (L2–L4)

A I IS

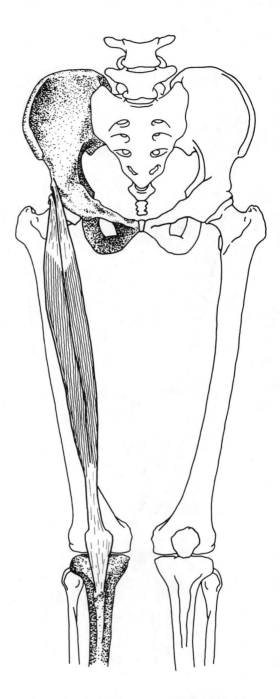

Hip, thigh, and leg—anterior view

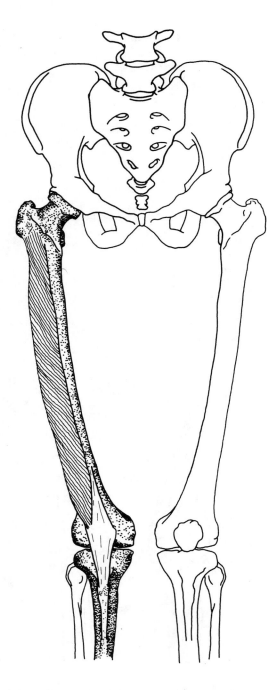

VASTUS LATERALIS
(One of quadriceps femoris)

Origin	Intertrochanteric line, inferior border of greater trochanter, gluteal tuberosity, lateral lip of linea aspera of femur
Insertion	Lateral margin of patella then by patellar ligament to tuberosity of tibia
Action	Extends leg at knee joint
Nerve	Femoral nerve (L2–L4)

Hip, thigh, and leg—anterior view

VASTUS MEDIALIS
(One of quadriceps femoris)

Origin   Intertrochanteric line, medial lip of
               linea aspera of femur, medial
               intermuscular septum, medial
               supracondylar line

Insertion Medial border of the patella then by
               patellar ligament into tibial tuberosity,
               medial condyle of tibia

Action   Extends leg at knee joint

Nerve   Femoral nerve (L2–L4)

Hip, thigh, and leg—anterior view

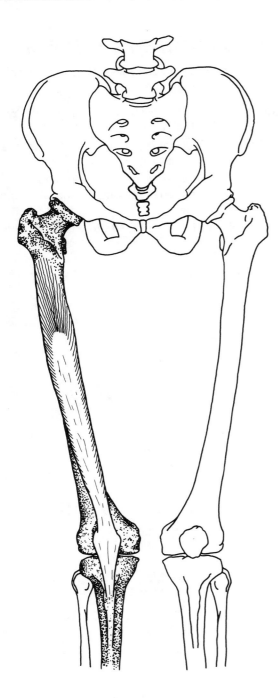

VASTUS INTERMEDIUS
(One of quadriceps femoris)

Origin	Anterior and lateral surfaces of upper two-thirds of femur, lateral intermuscular septum, linea aspera, and lateral supracondylar line
Insertion	Deep aspect of quadriceps tendon then through patella to tibial tuberosity
Action	Extends leg at knee joint
Nerve	Femoral nerve (L2–L4)

Hip, thigh, and leg—anterior view

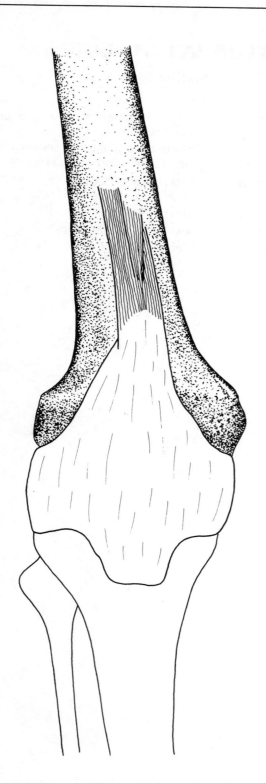

Knee—anterior view

ARTICULARIS GENUS

(May be considered part of vastus intermedius)

Origin	A few slips from front of lower part of femur
Insertion	Upper part of synovial membrane of knee joint
Action	Retracts synovial membrane superiorly during extension of knee joint
Nerve	Branch of nerve to vastus intermedius

BICEPS FEMORIS

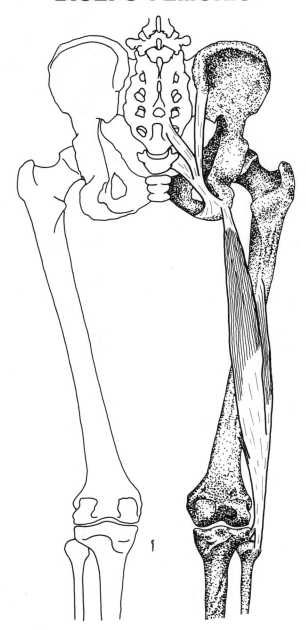

Hip and thigh—posterior view

Origin	Long head—ischial tuberosity, sacrotuberous ligament	**Action**	Flexes leg at knee joint, long head also extends thigh at hip joint
	Short head—linea aspera, lateral supracondylar ridge, lateral intermuscular septum	**Nerve**	Long head—tibial part of sciatic nerve (S1–S3)
Insertion	Lateral side of head of fibula and lateral condyle of tibia		Short head—common peroneal part of sciatic nerve (L5, S1, S2)

SEMITENDINOSUS

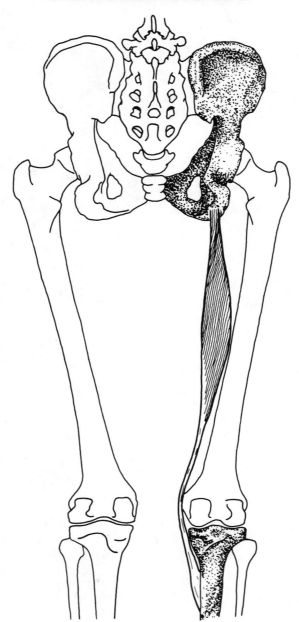

Hip and thigh—posterior view

Origin	Ischial tuberosity	**Action**	Flexes and slightly <u>medially rotates</u> leg at knee joint after flexion, extends thigh at hip joint
Insertion	Medial surface of shaft of tibia		
		Nerve	Tibial portion of sciatic nerve (L5, S1, S2)

SEMIMEMBRANOSUS

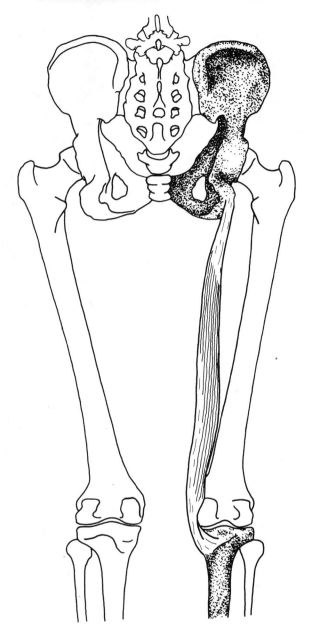

Hip and thigh—posterior view

Origin	Ischial tuberosity	**Action**	Flexes and slightly medially rotates leg at knee joint after flexion, extends thigh at hip joint
Insertion	Posterior part of medial condyle of tibia		
		Nerve	Tibial portion of sciatic nerve (L5, S1, S2)

GRACILIS

Origin   Lower margin of body and inferior ramus of pubis

Insertion   Upper part of medial surface of shaft of tibia

Action   Adducts thigh at hip joint and flexes leg at knee joint, with leg flexed it assists in medial rotation

Nerve   Obturator nerve (L3, L4)

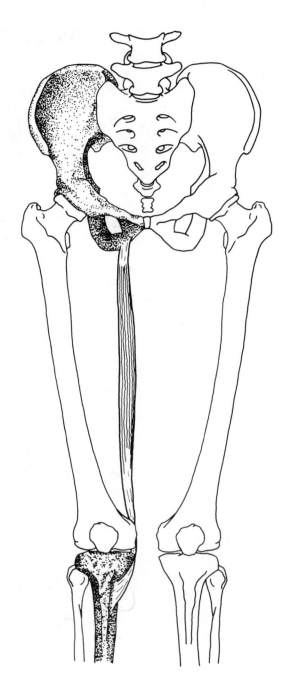

Hip and thigh—anterior view

PECTINEUS

Anterior Pubis

Origin	Pectineal line on superior ramus of pubis
Insertion	From lesser trochanter to linea aspera of femur
Action	Flexes and adducts thigh at hip joint, medially rotates thigh
Nerve	Femoral nerve (L2–L4), (sometimes a branch of obturator nerve)

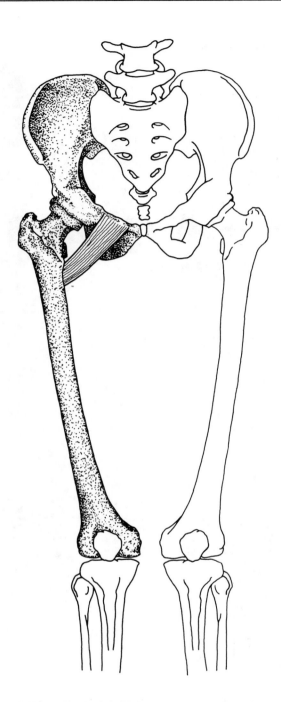

Hip and thigh—anterior view

ADDUCTOR LONGUS

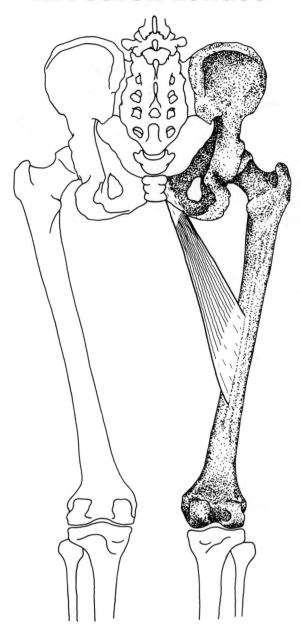

Hip and thigh—posterior view

Origin	Anterior of body of pubis	**Action**	Adducts thigh at hip joint, assists in lateral rotation
Insertion	Medial lip of linea aspera		
		Nerve	Obturator nerve (L3, L4)

ADDUCTOR BREVIS

Can not palpate

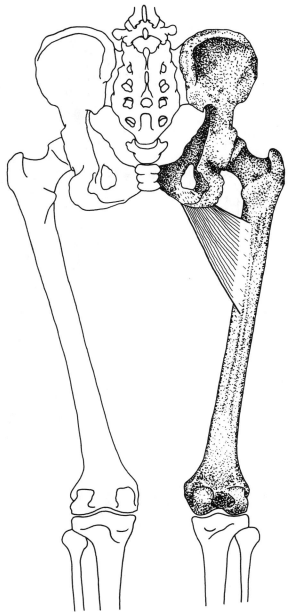

Hip and thigh—posterior view

Origin	Outer surface of inferior ramus of pubis	**Action**	Adducts thigh at hip joint, assists in lateral rotation
Insertion	From below lesser trochanter to linea aspera and into proximal part of linea aspera	**Nerve**	Obturator nerve (L3, L4)

ADDUCTOR MAGNUS

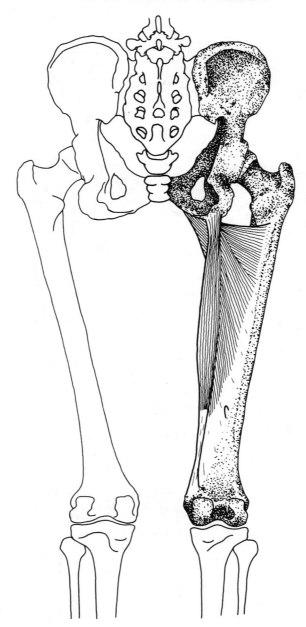

Hip and thigh—posterior view

Origin	Inferior ramus of pubis, and ramus and lower part of tuberosity of ischium	**Action**	Adducts thigh at hip joint, assists in lateral rotation and extension
Insertion	Linea aspera, adductor tubercle of femur	**Nerve**	Obturator nerve (L3, L4), sciatic nerve

ANTHIOR   post

CHAPTER NINE
MUSCLES OF THE LEG AND FOOT

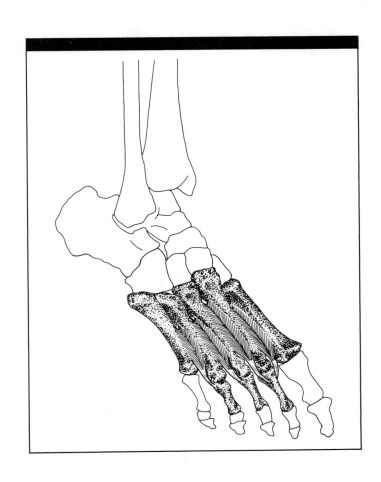

TIBIALIS ANTERIOR

Origin   Lateral condyle of tibia, upper half of
lateral surface of tibia, interosseous
membrane

Insertion   Medial side and plantar surface of
medial cuneiform bone, and base of *NAvicular*
first metatarsal bone

Action   Dorsiflexes foot at ankle joint, inverts
(supinates) foot

Nerve   Deep peroneal nerve (L4, L5, S1)

makes it pop out

if you do not have a drop foot Deep peroneal Nerve

Leg—anterolateral view

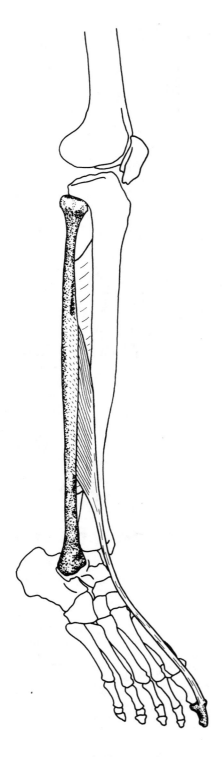

EXTENSOR HALLUCIS LONGUS

Origin	Middle half of anterior surface of fibula and interosseous membrane
Insertion	Base of distal phalanx of great toe
Action	Extends great toe, dorsiflexes and supinates foot
Nerve	Deep peroneal nerve (L4, L5, S1)

Leg—anterolateral view

EXTENSOR DIGITORUM LONGUS

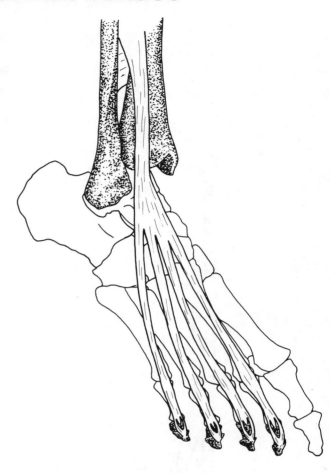

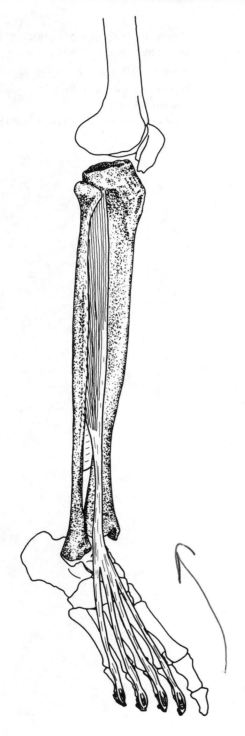

Foot—anterolateral view

Origin	Upper two-thirds of anterior surface of fibula, interosseous membrane, lateral condyle of tibia
Insertion	Into extensor expansion on dorsal surface of four lateral toes, and then to bases of middle and distal phalanges
Action	Extends toes, dorsiflexes foot at ankle
Nerve	Deep peroneal nerve (L4, L5, S1)

Leg—anterolateral view

PERONEUS TERTIUS

(Lower lateral part of extensor digitorum longus)

Origin	Lower third of anterior surface of fibula and interosseous membrane
Insertion	Dorsal surface of base of fifth metatarsal bone
Action	Dorsiflexes and everts foot
Nerve	Deep peroneal nerve (L4, L5, S1)

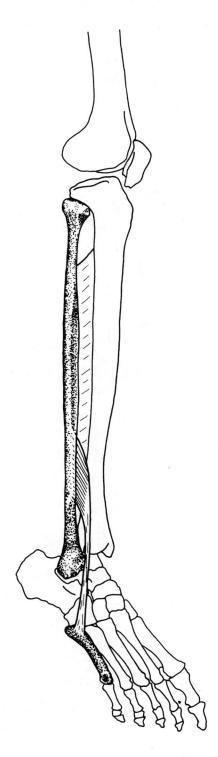

Leg—anterolateral view

GASTROCNEMIUS

Origin	Lateral head—lateral condyle and posterior surface of femur
	Medial head—popliteal surface of femur above medial condyle
Insertion	Posterior surface of the calcaneus
Action	Plantar flexes foot, flexes leg at knee
Nerve	Tibial nerve (S1, S2)

Leg—posterior view

SOLEUS

Origin	Posterior surface of the tibia (soleal line), upper third of posterior surface of fibula, fibrous arch between tibia and fibula
Insertion	Posterior surface of the calcaneus
Action	Plantar flexes foot
Nerve	Tibial nerve (S1, S2)

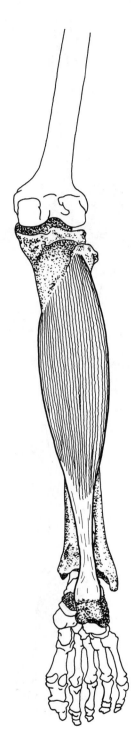

Leg—posterior view

PLANTARIS

Origin	Lateral supracondylar ridge of femur, oblique popliteal ligament
Insertion	Posterior surface of the calcaneus
Action	Plantar flexes foot, flexes leg
Nerve	Tibial nerve (L4, L5, S1)

Leg—posterior view

POPLITEUS

Origin	Lateral surface of lateral condyle of femur
Insertion	Upper part of posterior surface of tibia
Action	Rotates leg medially, flexes leg
Nerve	Tibial nerve (L4, L5, S1)

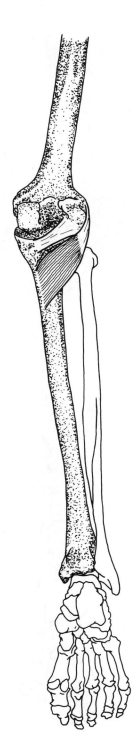

Leg—posterior view

FLEXOR HALLUCIS LONGUS

Origin	Lower two-thirds of posterior surface of shaft of fibula, posterior intermuscular septum, interosseous membrane
Insertion	Base of distal phalanx of great toe
Action	Flexes distal phalanx of great toe, assists in plantar flexing foot, inverts foot
Nerve	Tibial nerve (L5, S1, S2)

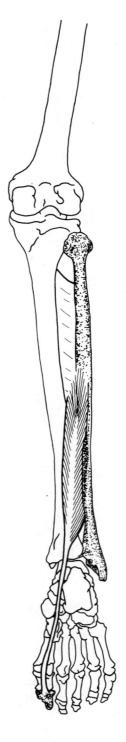

Leg—posterior view

FLEXOR DIGITORUM LONGUS

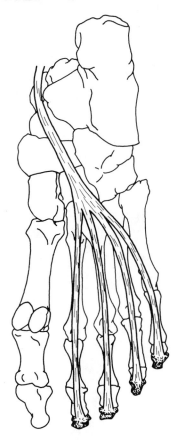

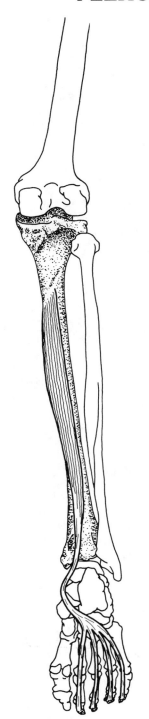

Foot—plantar view

Origin	Medial part of posterior surface of tibia
Insertion	Bases of distal phalanges of second, third, fourth, and fifth toes
Action	Flexes distal phalanges of lateral four toes, assists in plantar flexing foot, inverts foot
Nerve	Tibial nerve (L5, S1)

Leg—posterior view

TIBIALIS POSTERIOR

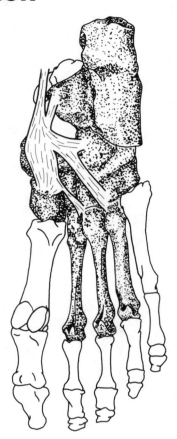

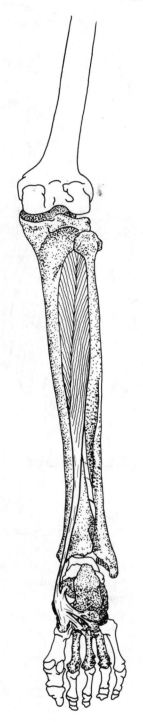

Foot—Plantar view

Origin	Lateral part of posterior surface of tibia, interosseous membrane, proximal half of posterior surface of fibula
Insertion	Tuberosity of navicular bone, cuboid, cuneiforms, second, third, and fourth metatarsals, sustentaculum tali of calcaneus
Action	Plantar flexes, inverts foot
Nerve	Tibial nerve (L5, S1)

Leg—posterior view

PERONEUS LONGUS

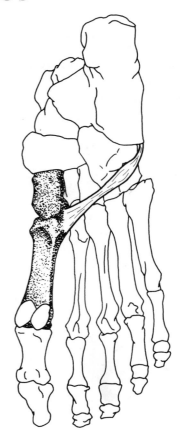

Foot—plantar view

Origin	Upper two-thirds of lateral surface of fibula
Insertion	Lateral side of medial cuneiform, base of first metatarsal
Action	Plantar flexes and everts foot
Nerve	Superficial peroneal nerve (L4, L5, S1)

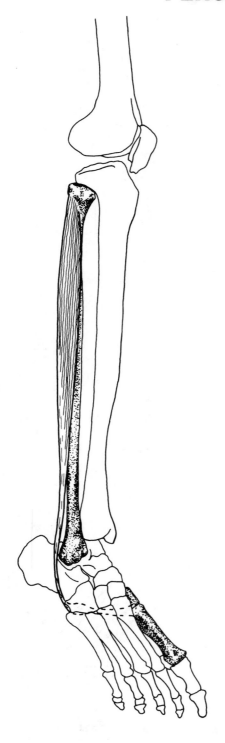

Leg—anterolateral view

PERONEUS BREVIS

Origin	Lower two-thirds of lateral surface of fibula
Insertion	Lateral side of base of fifth metatarsal bone
Action	Plantar flexes, everts foot
Nerve	Superficial peroneal nerve (L4, L5, S1)

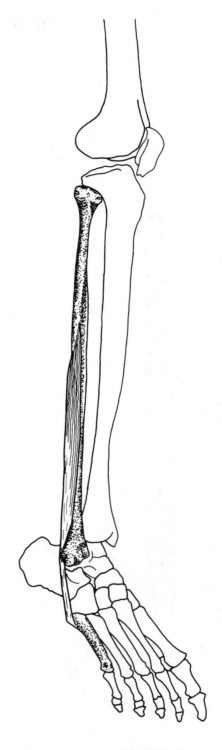

Leg—anterolateral view

EXTENSOR DIGITORUM BREVIS

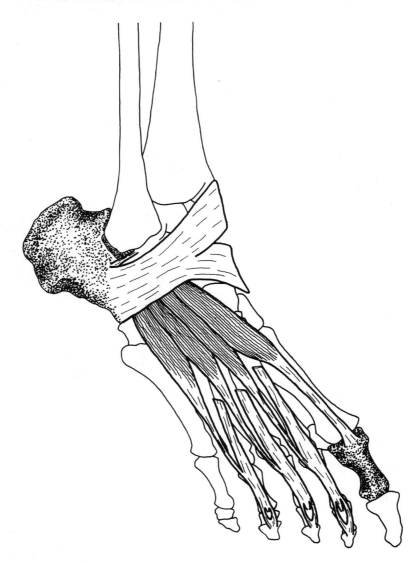

Foot—anterolateral view

Origin   Anterior and lateral surfaces of calcaneus, lateral talocalcaneal ligament, inferior extensor retinaculum

Insertion   Into base of proximal phalanx of great toe, into lateral sides of tendons of extensor digitorum longus of second, third, and fourth toes

Action   Extends the four toes

Nerve   Deep peroneal nerve (L5, S1)

ABDUCTOR HALLUCIS

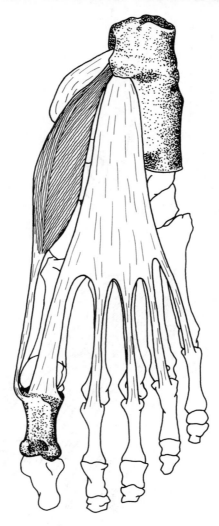

Foot—plantar view

Origin	Tuberosity of calcaneus, flexor retinaculum, plantar aponeurosis	**Action**	Abducts great toe
Insertion	Medial side of base of proximal phalanx of great toe	**Nerve**	Medial plantar nerve (L4, L5)

FLEXOR DIGITORUM BREVIS

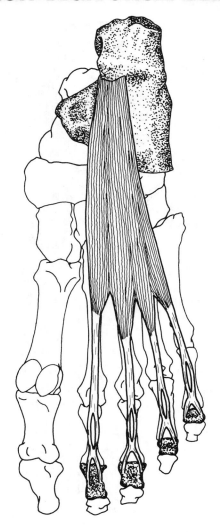

Foot—plantar view

Origin	Tuberosity of calcaneus, plantar aponeurosis	**Action**	Flexes proximal phalanges and extends distal phalanges of second through fifth toes
Insertion	Sides of middle phalanges of second to fifth toes	**Nerve**	Medial plantar nerve (L4, L5)

ABDUCTOR DIGITI MINIMI

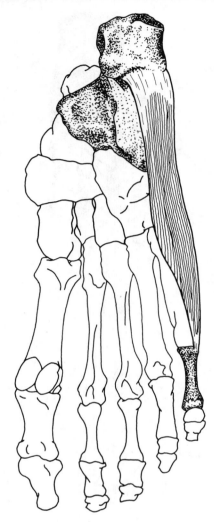

Foot—plantar view

Origin	Tuberosity of calcaneus, plantar aponeurosis	**Action**	Abducts fifth toe
Insertion	Lateral side of proximal phalanx of fifth toe	**Nerve**	Lateral plantar nerve (S1, S2)

QUADRATUS PLANTAE

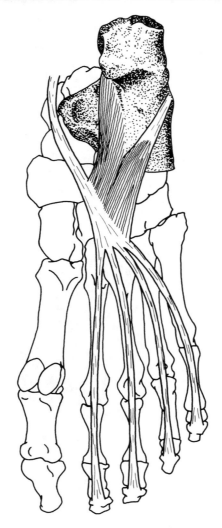

Foot—plantar view

Origin   Medial head—medial surface of calcaneus

Lateral head—lateral border of inferior surface of calcaneus

Insertion   Lateral margin of tendon of flexor digitorum longus

Action   Flexes terminal phalanges of second through fifth toes

Nerve   Lateral plantar nerve (S1, S2)

LUMBRICALES

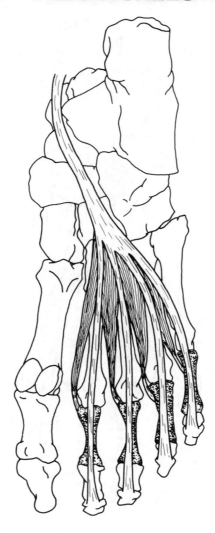

Foot—plantar view

Origin	Tendons of flexor digitorum longus	**Nerve**	First lumbricalis—medial plantar nerve (L4, L5)
Insertion	Dorsal surfaces of proximal phalanges		Second through fifth lumbricales—
Action	Flex proximal phalanges of second through fifth toes		lateral plantar nerve (S1, S2)

FLEXOR HALLUCIS BREVIS

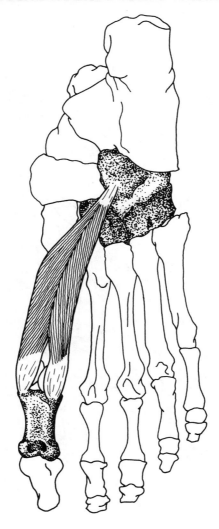

Foot—plantar view

Origin   Cuboid bone, lateral cuneiform bone

Insertion   Medial part—medial side of base of
proximal phalanx of great toe

Lateral part—lateral side of base of
proximal phalanx of great toe

Action   Flexes proximal phalanx of great toe

Nerve   Medial plantar nerve (L4, L5, S1)

ADDUCTOR HALLUCIS

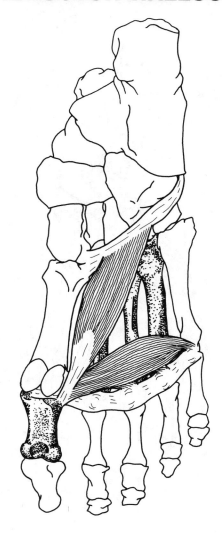

Foot—plantar view

Origin

Oblique head—second, third, and fourth metatarsal bones, and sheath of peroneus longus tendon

Transverse head—plantar metatarsophalangeal ligaments of third, fourth, and fifth toes, and transverse metatarsal ligaments

Insertion   Lateral side of base of proximal phalanx of great toe

Action   Adducts great toe

Nerve   Lateral plantar nerve (S1, S2)

FLEXOR DIGITI MINIMI BREVIS

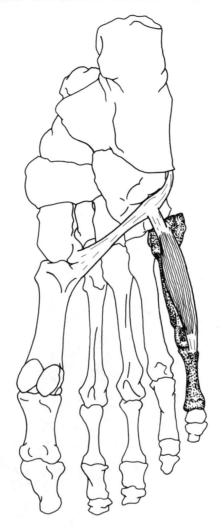

Foot—plantar view

Origin   Base of fifth metatarsal, sheath of
            peroneus longus tendon

Insertion Lateral side of base of proximal
            phalanx of fifth toe

Action   Flexes proximal phalanx of fifth toe

Nerve   Lateral plantar nerve (S1, S2)

DORSAL INTEROSSEI
(Four muscles)

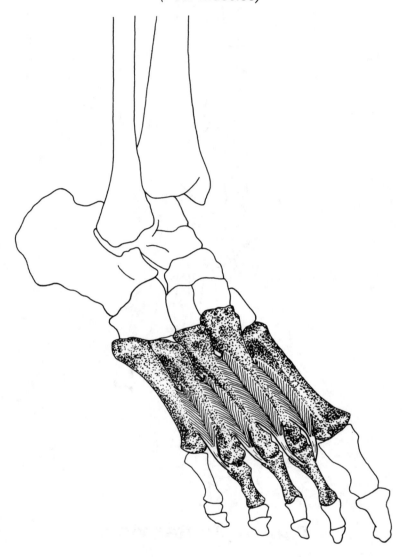

Foot—anterolateral view

Origin	Adjacent sides of metatarsal bones	**Action**	Abduct toes, flex proximal phalanges
Insertion	Bases of proximal phalanges	**Nerve**	Lateral plantar nerve (S1, S2)

Insertion
First—medial side of proximal phalanx of second toe

Second, third, fourth—lateral sides of proximal phalanges of second, third, and fourth toes

PLANTAR INTEROSSEI

(Three muscles)

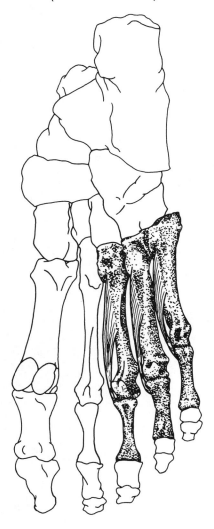

Foot—plantar view

Origin	Bases and medial sides of third, fourth, and fifth metatarsal bones	**Action**	Adduct toes, flex proximal phalanges
Insertion	Medial sides of bases of proximal phalanges of same toes	**Nerve**	Lateral plantar nerve (S1, S2)

ALPHABETICAL LISTING
OF MUSCLES

INDEX